Assessment of Power System Reliability

Marko Čepin

Assessment of Power System Reliability

Methods and Applications

Springer

Prof. Dr. Marko Čepin
Faculty of Electrical Engineering
University of Ljubljana
Trzaska 25
1000 Ljubljana
Slovenia
e-mail: marko.cepin@fe.uni-lj.si

ISBN 978-1-4471-6100-4 ISBN 978-0-85729-688-7 (eBook)
DOI 10.1007/978-0-85729-688-7
Springer London Dordrecht Heidelberg New York

British Library Cataloguing in Publication Data
A catalogue record for this book is available from the British Library

Cover design: eStudio Calamar S.L.

Printed on acid-free paper

Springer is part of Springer Science+Business Media (www.springer.com)

To Alenka, Katarina and Irena

Preface

The importance of power system reliability, which is the subject of this book, is demonstrated, when people are confronted with the loss of electrical energy. This is true, no matter if the loss causes stopping of the production lines or even the shutdown of our companies, which can consequently cause huge economic deficits, or the loss of electrical energy only slightly decreases the comfort of our free time in our homes. Fortunately, such events happen rarely, because the reliability of modern power systems is high.

Our aim for the reliability is connected to its constant improvement as this is in relation to the progress of the modern world. The objective of this book is to contribute to such improvement.

The book is divided to six parts, which comprise 20 chapters.

Part I of this book is related to introductory chapters comprising the background issues important to power system reliability. Chapter 1 briefly touches the history of power systems, which evolved from early days of discovering the electrical energy to today, and which is expected to be expanded in the future with new technologies, new features and new application fields. Chapter 2 presents short writing about selected blackouts, which are some of the main issues with which the power system reliability has to face. As the complexity of the systems increases in general, it is expected that the undesired events such as blackouts may result in larger consequences. Chapter 3 gives the main definitions of reliability and connected parameters. The largest difficulty of these definitions is that their use and understanding through different technical disciplines is not always the same. Chapter 4 summarizes the probability theory, which represents the background for reliability calculations. The probability theory and its implications can be very complex, but only their features, which facilitate the reliability evaluations of power systems, are emphasized inside this book.

Part II of this book is related to reliability methods, which can be used for analyses of technical systems and processes. Chapter 5 presents the fault tree analysis, and the procedure for its practical applications is given. The identification and evaluation of important factors are presented. Chapter 6 shortly describes the event tree analysis, which is a method for evaluation of scenarios. Chapter 7 presents the binary decision diagram, which is a directed graph that consists of nodes and edges and that deals with Boolean logic. Chapter 8 gives the Markov processes, where the states of the components and the systems and the transitions between them are evaluated. Chapter 9 describes the reliability block diagram, which is the least abstract of the reliability methods. Chapter 10 deals with common cause failures, which represent the evaluation of dependent events and are important for consideration in redundant systems.

Part III of this book is related to the methods of the power flow analysis, because the dynamic aspect of the power system is an important part of the related reliability assessments. Chapter 11 includes brief overview of the selected power flow methods.

Part IV is the most important part of this book. It is related to various aspects of assessing the reliability of the power system and its parts. Chapter 12 presents the generating capacity methods including practical examples. Chapter 13 lists the reliability and performance parameters of selected power plants, which gives information about their successful operation from various points of view. Chapter 14 is oriented to distribution system reliability measures. This chapter may be understood as the most important chapter of this part, because the larger number of difficulties with power system reliability is connected with distribution systems. Chapter 15 gives a power system reliability method, which in addition to its static feature of assessing the static power system configuration includes the dynamics of the system in a certain way. The extent of power flows through the portions of the system that may impact the calculation of the power system reliability.

Part V of this book is related to the optimization methods. The field of optimization methods is very wide and many methods exist. Only the selected methods among those that may be the most applicable for the tasks connected with reliability improvement are mentioned. Chapter 16 shortly presents the linear programming, which is among the simplest methods aimed to solve problems in terms of linear equalities and inequalities. Chapter 17 gives the dynamic programming and its main feature is that it transforms a complex problem into a sequence of simpler problems. Chapter 18 summarizes the genetic algorithms, which is a step toward solving nonlinear problems. Chapter 19 describes briefly the simulated annealing, which is another method for solving nonlinear problems, where the local minima should be avoided when finding the global one.

Part VI of this book is related to the application of reliability and optimization methods. Chapter 20 represents a wide field with many reliability applications and optimizations, from which only some selected ones are presented.

The index at the end includes the most important terms related to the subjects of the book and identifies their placement in the book.

The chapters of the book have been aimed to be written in a simple language without very detailed theoretical and mathematical abstraction of specific methods and their applications. Rather, the practical use of the materials directed the writing of the book. This book contains simple examples in most of the chapters or within their sections in order to facilitate the understanding of the theory behind.

Each chapter includes a list of references, which support the content of the chapter. The Internet addresses of some references, which are available at the time of writing the book, are provided. In addition, several references may exist on the Internet, which were not available at the time of writing of the book, or the Internet address of existing information may change, so the readers of this book are encouraged to search the Internet for the supporting references of their interest. It is suggested that the search is made by putting the title of the reference in brackets in order to reduce the list of unnecessary results, e.g. search of "component reliability data for use in probabilistic safety assessment", which will give a link to the available full document on the subject.

The expected users of this book are the power engineering students, both at undergraduate and at graduate level, and the engineers in the electric power industry.

Many issues related to the reliability of power systems are not included here, because it is not possible to include all the issues of such important and wide field, as it is the power system reliability, in one book.

Ljubljana, 10 January 2011 Marko Čepin

Acknowledgments

It is very difficult to judge, which of the authors of a scientific work contributed more and which less or to assess which contribution is more valuable. Similarly, it is very difficult to give the recognition they deserve to all the people, who contributed to this book.

In certain way, the following colleagues, whose contributions are the most notable, are emphasized.

Dr. Andrija Volkanovski helped in discussions about the variants of the power system reliability method and their applications. He developed the power system reliability method as a part of his PhD under my supervision. He is a co-author of Chap. 15, Power system reliability method.

The final version of this book has been written much better because of the independent reviewers, who reviewed the writings and contributed with helpful comments and suggestions, which improved the contents of the book.

The list of independent reviewers includes: Prof. Dr. Rafael Mihalič, Prof. Dr. Iztok Tiselj, Prof. Dr. Igor Papič, Prof. Dr. Miloš Pantoš, Prof. Dr. Ferdinand Gubina, Asst. Prof. Dr. Ivo Kljenak, Prof. Dr. Tomaž Gyergyek, Dr. Andrej Prošek, Dr. Matjaž Leskovar, Dr. Mitja Uršič, Blaže Gjorgiev, Duško Kančev, Martin Draksler, Blaž Likovič and Luka Trček.

Selected specific parts of specific chapters of this book were prepared with the help of my colleagues and postgraduate students.

Duško Kančev helped in the preparation of sections about power flow methods. Blaže Gjorgiev helped in the preparation of sections about genetic algorithms and simulated annealing. Marko Kolenc helped in the preparation of sections about generating capacity methods. Živa Bricman helped in the preparation of section about blackouts, section about reliability measures and section about event trees. James Cornwell helped in the preparation of section about binary decision diagrams, section about Markov models and section about reliability block diagrams. Gregor Praznik helped in the preparation of section about reliability indices of distribution systems.

I thank sincerely to all the listed colleagues for their contributions and efforts, which helped me in preparation of this book.

I especially thank my own family, Alenka, Katarina and Irena for their understanding and constant encouragement.

Contents

Part I Background

Part II Reliability Methods

Part I
Background

Chapter 1
Introduction to Power Systems

Things should be as simple as possible, but not simpler
Albert Einstein

1.1 Introduction

Electric power system is one of the largest and the most complex systems established by the mankind. It consists of uncounted number of facilities and structures, systems and subsystems, components and equipment, and the complex interactions among all those.

Those interactions include physical interactions, operational interactions, and administrative interactions. Physical interactions are interaction between parts of the facilities, systems, and components that are physically connected. Operational interactions are the interactions where the equipment electrically, magnetically, or mechanically interacts with other equipment. Administrative interactions are the interactions where the facilities and structures, systems and subsystems, components and equipment are subjected to management and administrative procedures, which include a wide set of documents. The examples of the related documents include the general operating procedures, abnormal operating procedures, emergency operating procedures, operating manual, technical specifications, management guidelines, and system description documents, including testing and maintenance requirements and corresponding procedures.

The benefits of electric power systems are integrated into the much faster modern life in such extent that it is impossible to imagine the society without the electrical energy as the main output of those systems. This emphasizes the importance of power systems' reliability, which is the subject of this book. The beginning of understanding the importance of the power systems' reliability has not been initiated so far in the history.

M. Čepin, *Assessment of Power System Reliability*,
DOI: 10.1007/978-0-85729-688-7_1, © Springer-Verlag London Limited 2011

1.2 Short History of Electric Power Systems

The history of electric power systems started more than two centuries ago, and it is not limited with the names of Benjamin Franklin, Alessandro Volta, Luigi Galvani, Hans Christian Örsted, Michael Faraday, André-Marie Ampère, Werner Siemens, Nikola Tesla, Thomas Alva Edison, and George Westinghouse. They are mentioned as they have been identified as important persons with their significant discoveries, which contributed to our electric power systems [1–11].

1.2.1 People

Benjamin Franklin was one of the first people in early eighteenth century who admired lightning during thunderstorms [9]. He invented lightening conductor and discovered positive and negative electrical charges, which he called fluid at that time. Alessandro Volta published a description of a silver–zinc battery in the year 1800 as a series of two different metal disks separated by a cardboard disk soaked with acid or salt solutions, but he did not know how to make it work. Luigi Galvani discovered a galvanic cell, which consisted of two different metallic slip sunken in a salty solution, as he accidently touched dead frog's leg with two different metals [10]. Hans Christian Örsted discovered the connection between electric and magnetic field in the year 1820. He ascertained that an electric current causes a compass needle to change its direction. A year later, Michael Faraday invented a simple electric generator, so-called Faraday disk. He built a machine that could convert mechanical power into electric.

André-Marie Ampére was interested in electromagnetic field. He proved that a coil containing electric current behaves similar to a stick magnet and that it creates magnetic field. Werner Siemens is known by perfecting the dynamo. Dynamo is a generator in which a part of its current is used to power the field windings, eliminating the usage of permanent magnets.

Thomas Alva Edison founded a practical incandescent light in the year 1879, which replaced arc lamps. He also developed the main idea for a complete distribution system with underground cables, electric wiring, fuses, and switches. Edison was enraptured with direct current. George Westinghouse was responsible for the introduction and development of alternating current for light and power (http://www.westinghousenuclear.com/Our_Company/history/george_westinghouse.shtm). Westinghouse also installed the first hydro power plant at Niagara Falls designed by Nikola Tesla.

Nikola Tesla was definitely one of the most significant men in electric power history. His inventions included alternating current, polyphase electric motor, and rotating magnetic field.

1.2.2 Early Power Transmission

Various systems have been used for power transmission. Mostly used options were the following:

- Telodynamic transmission, which means cable in motion, was mostly used for cable cars and street cars, as their electricity demand was quite high.
- Pneumatic transmission, which means transmission with compressed air, was used at the beginning of twentieth century in Paris, Birmingham, Offenbach, Dresden, and Buenos Aires.
- Hydraulic transmission, which means transmission using water under high pressure, was used to deliver power to factory motors.

Much of the early power transmission was with direct current. In early ages, power transmission faced two main obstacles: (1) the difficulty of voltage adjustment and (2) long-distance transmission. Therefore, companies used different lines for different voltage levels. As there were so many specific lines and the transmission was so inefficient, that generators needed to be close to their loads, it seemed that electric industry would develop into a distributed generation system with large numbers of small generators located nearby their loads. The early direct current power plants needed to be installed within 2.4 km of the farthest consumer.

Direct current system needed thick cables, and the losses of direct current cables were huge as they are proportional to the square of the current, the length of the cable, and the resistivity of the material. Early transmission networks used copper, which was one of the best economically feasible materials.

1.2.3 Early High-Voltage Systems

The idea of building a central plant and a network to deliver energy to customers, who pay fee for service, was a significant business model for inventors. The San Francisco Power System started to sell electrical energy from a central power plant to consumers through transmission lines in year 1879. A central power station to supply 3-km length of arc lighting in Broadway was built in 1880. Many other US cities produced public light by the end of 1881. Berlin had high-voltage arc lights for stores and public areas from 1884. Large amounts of power became reachable after 1889 when Charles Parson began to produce turbo generators. Generators output quickly jumped from 0.1 to 25 MW of electrical power.

1.2.4 Alternating Current

George Westinghouse thought that low voltage of direct current was too inefficient to be suitable for large system transmission. He concluded that long-distance transmission needed higher voltage, which is less expensive only with alternating current.

Between years 1884 and 1885, Hungarian engineers Zipernowsky, Bláthy, and Déri created closed core coil. They discovered that all former coreless or open-core devices were incapable of voltage control. Their patents were known as "closed-core transformer" and "shell-core transformer." In both transformers, the magnetic flux links the primary and the secondary coils and travels almost entirely within the iron core. With these inventions, it was technically and economically possible to provide electric power to households, businesses, and public spaces.

Nikola Tesla developed an equipment that allowed generation and use of alternating current in year 1888. Westinghouse bought patents for Tesla step-up transformer, polyphase alternating current generator, and motor design, which are still in use today. Three-phase generator and motor invented by Tesla were more efficient, cheaply manufactured, and more compact as existing direct current dynamos. The alternating current equipment proposed by Tesla prevailed over the direct current equipment proposed by Edison also because of the lower cost of step-up and step-down transformers.

A high-voltage three-phase transmission of alternating current was realized in 1891 in Frankfurt. A transmission line of 25 kV and 175-km long was built between Lauffen and Frankfurt. Transmission lines were supported by porcelain pin-and-sleeve insulators, which had a limit of 40 kV. The invention of disc insulator in the year 1907 allowed practical insulators of any length to be constructed for higher voltages.

1.2.5 Modern Period

The first transmission line in North America was constructed in 1889 between Oregon City and Portland with the length of 21 km and operating at 4 kV. There were 55 transmission systems operating at more than 70 kV installed by the year 1914. The first 110-kV line in Europe was built in the year 1912 in Germany, and the first 220-kV line was built in San Francisco Bay area in early 1920s. Electrical grid was largely expanded during World War I when large power-generating plants were built by governments and later transmission lines connected them to consumers.

The electrical networks in UK became interconnected in the national grid in the year 1926 initially operating at 132 kV. The first 220-kV line in Europe was built in the year 1929 between Cologne and Ludwigsburg-Hoheneck in Germany. This was one of the world largest power systems at that time. These lines were later upgraded to 380 kV.

The first 380-kV line was built in Sweden in the year 1952. The 952-km long line connected Harsprånget and Hallsperg. The first extra high-voltage transmission line at 735 kV linked the Manic–Outardes generating stations to the metropolitan areas of Québec and Montréal in the year 1965. The first transmission line at 1,200 kV was built in the year 1982.

Today, alternating current transmission is still used in distribution because of economy, efficiency, and reliability of transformers. Direct current is used for transmission of large amounts of power over long distances, or for interconnecting neighbor alternating current networks, mostly under the sea, but not for distribution.

1.3 Consumption and Complexity of Power Systems

The evolution of power systems goes hand in hand with the increase of generation and consumption of electrical energy and with the increase of systems complexity. The consumption of electrical energy has been increasing from its invention with the exception of the last world economic crisis around the year 2009. Figure 1.1 shows an example of electrical energy consumption in the period of 27 years for the selected regions: Europe, North America, Asia, and Oceania and for the global world consumption [12].

Figure 1.2 shows increase of electrical energy consumption for those regions and for the whole world. Yearly increases and cumulative increases, where the year 1980 is considered as a starting point, are shown separately.

Figures confirm the known fact that the consumption of electric energy increases every year for a couple of percents. The yearly increase of electric energy consumption is around 3% worldwide, around 2–3% in Europe and in North America, and it is slightly higher in Asia and Oceania, where the average yearly increase of electric energy consumption during the last 27 years stays around 6%. Those numbers differ from country to country and are higher in countries that develop faster. The world economic crisis has stopped the increase of electric energy consumption for a short period of time, but on the long term, the consumption is going to increase further.

Similarly, it can be stated for the complexity of power systems, although the complexity increases even faster. For example, if the number of power stations in the power system increases somehow linearly, the number of their cross-connections increases much faster.

The power systems are more and more dispersed and more and more integrated among themselves. They are more and more integrated with other systems as well, such as telecommunication and information systems.

The increase of complexity is not only physical. New and improved demands for quality of electric energy and for clean and safe environment are generated, which in parallel require more support systems and more interaction between them [33–37].

In addition, new systems are invented, which open completely new dimensions connected with power systems complexity. Only three topics are mentioned here to highlight the new dimensions of possible future expansion: (1) electric automobiles and other electric means of transport, (2) smart grids, and (3) international thermonuclear experimental reactor (ITER).

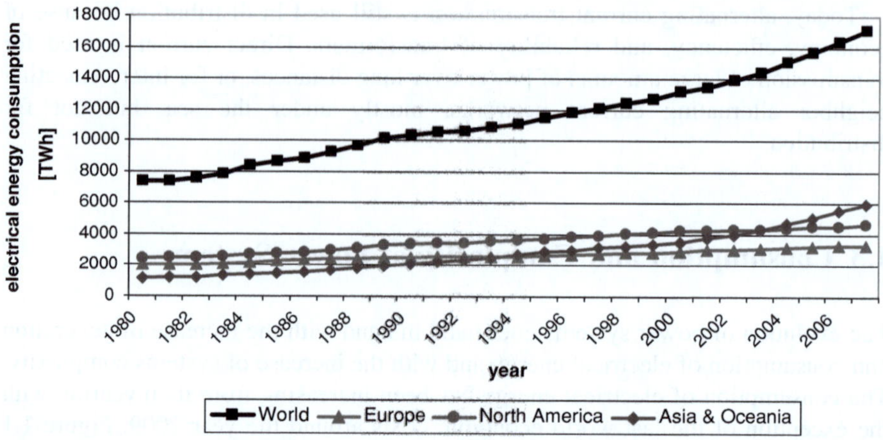

Fig. 1.1 Electrical energy consumption (selected years: 1980–2007)

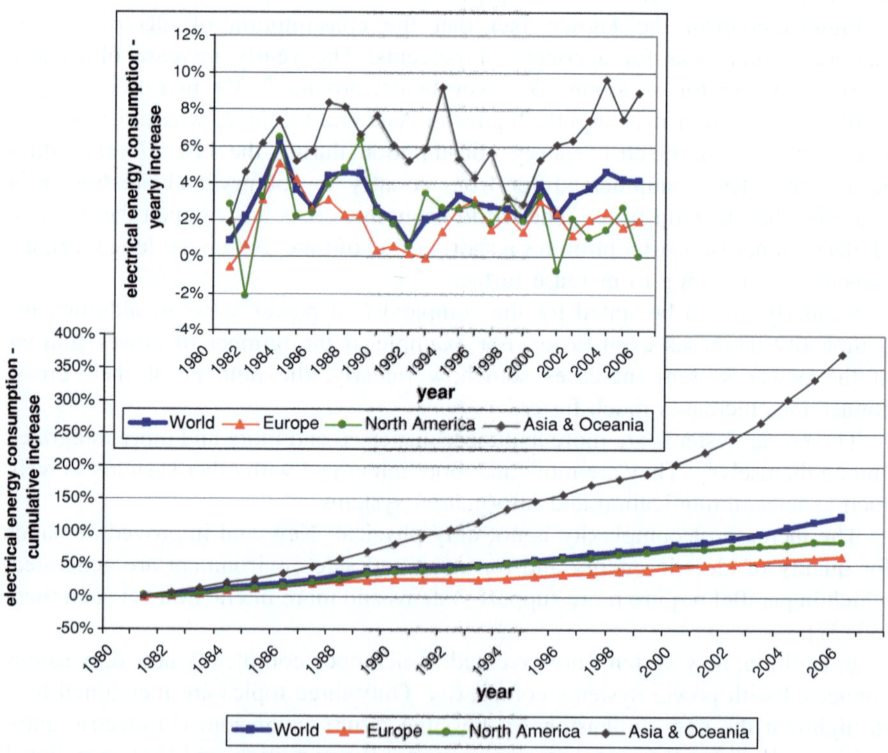

Fig. 1.2 Increase of electrical energy consumption (selected years: 1980–2007)

1.3.1 Electric Automobiles and Other Electric Means of Transport

Electric automobiles and other electric means of transport have been a negligible parameter of consumption within the power systems in the past. In the near future, they are going to be an important parameter of consumption of electric energy and at the same time an important parameter of electric energy storage.

1.3.2 Smart Grids

Smart grids (or SmartGrids, as sometimes related as) is a term that integrates the existing and new features of power systems into a modern power system with a number of abilities including at least the following [13–16].

- To ensure secure and sustainable electrical energy supplies and to combine the primary resources of traditional energy sources, flexible storage, and new and dispersed generation sources.
- To increase the network and generation capacity and to develop technical solutions that can be deployed rapidly and cost-effectively, enabling existing grids to accept power injections from distributed energy resources without exceeding operational limits.
- To establish interfacing capabilities that allow new designs of grid equipment and new automation and control arrangements to be successfully interfaced with existing traditional grid equipment.

The term smart grids relates to the modernization and optimization of the power grid. This means that the power grid will be more environmentally friendly, more efficient, and more reliable and secure. The establishment of smart grid requires integration and collaboration between several disciplines, including power systems, information systems, computer systems, and telecommunications systems.

The prerequisite, which goes along with the technical developments, is at least establishment of technical standards and protocols and development of regulatory and commercial frameworks in individual countries and their harmonization to facilitate cross-border trading. The essential feature of smart grids is establishment of technical standards and protocols that ensure open access, enabling the deployment of equipment from any chosen manufacturer without fear of lock-into proprietary specifications, which should apply to grid equipment, metering systems, and control and automation architectures.

1.3.3 International Thermonuclear Experimental Reactor

International thermonuclear experimental reactor (ITER—Latin world for "the way") is an international experiment aimed at demonstrating the technological feasibility of fusion energy. If the technological feasibility of fusion will be demonstrated within international thermonuclear experimental reactor, which will not be used for production of electric energy, the demonstration power plant may be built and a new era of complex power systems will begin [17–19]. Theoretical description of fusion power plant starts with heavy forms of hydrogen. When heavy forms of hydrogen (deuterium and tritium) are joined together at high temperature, the production of heat energy is achieved. When deuterium and tritium fuse, they form a helium nucleus, which is an alpha particle, and a high-energy neutron. In a fusion power plant, the energy of motion of the high-speed neutrons produced in the fusion reactions is converted to heat to make steam for the generation of electrical energy. The high temperature, which the fuel must hold, is a temperature of more than 100 million degrees Celsius. At such high temperature, the electrons are detached from the nuclei of the atoms, in a state of matter called plasma. The magnetic fields are used to confine the plasma. Comparing the effectiveness of other energy sources with the fusion regarding the mass of the fuel, the deuterium–tritium fusion process releases approximately three times as much energy as fission of uranium 235, and millions of times more energy than the burning of coal, which is a chemical reaction. The goal of a fusion power plant is to harness this energy to produce electrical energy. In theory, even other light elements and their isotopes are similarly useful as deuterium and tritium, but the latter are far better in practice as they require the lowest activation energy and thus the lowest temperature [17–19].

All mentioned in this subsection confirms the current and future complexity of power systems.

Because of the complexity of the electric power systems, it is relatively difficult to define and assess the reliability as a single parameter of a single system. Rather, several methods, tools, and measures are developed, which highlight the questions about the power systems reliability each from its viewpoint.

1.4 Objectives of the Book

The objective of this book is to present selected methods and measures, which can contribute to the answers about questions connected with the reliability of the electric power systems. The selection of the methods and measures is carried out in a way to give the reader of the book the best possible perspectives to understand the problems connected with the reliability of the power system and to suggest the most appropriate ways of getting the desirable answers.

The field of interest is too wide that all the information is captured within the pages of the book, so the references are carefully selected to lead the reader of the book to obtain more information in other books, papers, reports, and Internet pages.

1.5 Problems of Definitions of Reliability

One of the first and very important questions is the definition of the reliability of the power system itself [20–24]. The importance of this definition is so high mainly because of the fact that the reliability is a technical field, which is relatively new and which is used in many technical areas, where not a lot of mutual communication exists. Relative isolation of defining the objectives, stating the problems, and getting the solutions among the fields and disciplines has led to the fact that there are certain differences in basic views to the definition of the reliability, differences at identifications of problems, and differences in ways of solving them.

In some cases, even within one technical field, the experts are not of a unique opinion about the definition of reliability and application of its definition in a common way.

For example, in the software community, there are two groups of people. One group neglects the definition of software reliability as quantitative term, because they claim that software always behaves exactly as it was written and its reliability is by definition not needed at all. The other group looks to the definition of software reliability from a view point of software user. They understand the software reliability as quantitative term, which shows how the software behaves versus what is expected from it. The complete software life cycle is under the evaluation, but the requirements specification of the software is the crucial phase, which directs the software reliability. In other words, the software reliability is defined to cover the discrepancy between what is expected from the software and what is achieved by the software [25–27].

In some cases, the term reliability can be related to a product or to a process. If it is related to a product, the static perspective of reliability is much simpler defined. If it is related to the process, both the static perspective of reliability and the dynamic aspect of reliability can be of interest. This is the case in the power systems, where the term reliability is divided to adequacy and security [22]. The adequacy is related to the existence of sufficient generation of the electric power system to satisfy the consumer demand. The security is related to the ability of the electric power system to respond to transients and disturbances that occur in the system.

In some cases, the term reliability can be related only to a specific set of functions and not to all functions of the system. In such a case, it is difficult to distinguish what should be considered and what not.

In some cases, the term reliability is related to a process, which is managed by people, and the reliability of human operators, organizers, and managers plays an important role within the reliability of the system. The sociological and psychological aspects play role in addition to technological parameters in such cases [28–32].

1.6 Improvement of Reliability of Power Systems

The performance of the distribution system determines greater than 80% of the reliability of service to customers. The high-voltage transmission and generation system determine the rest. If the reliability is to be improved, the focus should be placed on the distribution system [24], although also the transmission and generation should not be taken out of considerations.

This reminds us about the 80–20 rule or the Pareto principle or the vital few law. They all mean the same: approximately, 80% of effects come from 20% of causes. The related applications have been found in different disciplines but they come to the same general conclusion. The rule originates from owning a land in Italy, where Pareto found out that 80% of Italian territory is owned by 20% of population. This sentence can be read in opposite way that 20% of the Italian territory was owned by 80% of people. The principle was widely applied in various disciplines and areas:

- *Economy*: 80% of the property and income belong to 20% of the richest people.
- *Computer science*: 80% of computer crashes are caused by 20% of bugs, or 80% of the use of commercial software base on 20% of the installed features.
- *Healthcare*: 80% of resources are used by 20% of patients.
- *Work*: 80% of your results are produced using 20% of your efforts.

References

1. Kirby RS, Withington S, Darling AB, Kilgour FG (1956) Engineering in history. McGraw-Hill, New York
2. Arrillag J (1998) High voltage direct current transmission. The Institution of Electrical Engineers, London
3. Thury R (2006) http://www.electrosuisse.ch/cms.cfm/s_page/74440. Accessed 27 July 2010
4. Hughes TP (1983) Network of power: electrification in Western society (1880–1930). John Hopkins University Press, Baltimore
5. Brown MH, Sedano RP (2004) Electricity transmission: a primer. National Council on Electric Policy, Washington, DC
6. Foran J, The day they turned the falls on: the invention of the universal electric power system. http://ublib.buffalo.edu/libraries/projects/cases/niagara.htm. Accessed 2 Sep 2010
7. Zipernowsky K, Déri M, Bláthy O T (1886) Induction-coil. United States patent office
8. George Westinghouse. http://www.westinghousenuclear.com/Our_Company/history/george_westinghouse.shtm Accessed 27 July 2010
9. Benjamin Franklin: Glimpses of the Man. http://www.fi.edu/franklin/ Accessed July 27 2010
10. Luigi Galvani. http://www.corrosion-doctors.org/Biographies/GalvaniBio.htm Accessed July 27 2010
11. Power Standards Lab, Early history of electric power. http://www.powerstandards.com/FunStuff/EarlyHistory/ElectricPower.htm. Accessed 27 July 2010
12. US Energy Information Administration (2008) http://www.eia.doe.gov. Accessed 27 July 2010
13. European Technology Platform SmartGrids (2006) Vision and strategy for Europe's electricity networks of the future. EUR 22040, EU

14. DOE (2008) The smart grid: an introduction. http://www.oe.energy.gov/DocumentsandMedia/DOE_SG_Book_Single_Pages(1).pdf. Acessed 27 July 2010
15. http://smartgrid.ieee.org/. Accessed 27 July 2010
16. SmartGrids: European Technology Platform. www.smartgrids.eu. Accessed 27 July 2010
17. ITER. http://www.iter.org/. Accessed 27 July 2010
18. ITER and the promise of fusion energy (2006) http://www.pppl.gov/projects/pics/ITER4pg.pdf. Accessed 27 July 2010
19. The US and ITER: the path to fusion energy. http://www.pppl.gov/projects/pics/ITER_DOE_OffSci.pdf. Accessed 27 July 2010
20. Reiche H (1972) Reliability definitions. Microel Rel 11(5):425–427
21. Dummer GWA, Tooley MH, Winton RC (1997) An elementary guide to reliability. Elsevier, San Diego, CA
22. Billinton R, Allan R (1996) Reliability evaluation of power systems. Plenum Press, New York
23. Brown RE (2002) Electric power distribution reliability. CRC Press, Boca Raton
24. Short TA (2006) Distribution reliability and power quality. Taylor & Francis, LLC
25. Laprie JC (1996) Software-based critical systems. In: Proceedings of SAFECOMP96, Vienna, pp 157–170
26. Čepin M (1998) Improving reliability of computerised safety-related systems. PhD thesis, University of Ljubljana
27. Čepin M, Mavko B (1999) Fault tree developed by an object-based method improves requirements specification for safety-related systems. Rel Eng Syst Saf 63:111–125
28. Swain AD, Guttmann HE (1983) Handbook for human reliability analysis with emphasis on nuclear power plants application. NUREG/CR-1278 (SAND80-0200), NRC
29. Čepin M (2009) IJS-HRA: a method for human reliability analysis. Asigurarea Calitatii 15(57):21–27
30. Čepin M (2008) DEPEND-HRA: a method for consideration of dependency in human reliability analysis. Rel Eng Syst Saf 93(10):1452-1460
31. Prošek A, Čepin M (2008) Success criteria time windows of operator actions using RELAP5/MOD33 within human reliability analysis. J Loss Prev Proc Ind 21(3):260–267
32. Čepin M (2008) Importance of human contribution within the human reliability analysis (IJS-HRA). J Loss Prev Proc Ind 21(3):268–276
33. Smith W (2009) Electric power system reliability. Powersmiths
34. Philipson L, Lee Willis H (2006) Understanding electric utilities and de-regulation. Taylor & Francis, New York
35. Cazassa JA (1994) The development of electric power transmission. IEEE Press, New York
36. Jonnes J (2003) Empires of light. Random House, New York
37. Willis HL, Schreiber RR, Welch GV (2001) Aging power delivery infrastructures. Marcel Dekker, New York

Chapter 2
Introduction to Blackouts

Life is really simple, but men insist on making it complicated
Confucius

2.1 Introduction

Power system blackout means that a larger area of consumers of electrical energy is left without electrical energy for a determined duration of time [1–3]. This area of consumers can include the suppliers of electrical energy, which because of the blackout stop providing the electrical energy if the conditions require so. Power system brownout is a temporary interruption of power service in which the electric power is reduced [4].

Power system blackouts have become a phenomenon seems to be more important. The importance of blackouts is large, because their economical and technical consequences may grow up. Larger and larger consequences are mostly connected with the increased complexity of the systems. This requires the increase of the number of activities in the field of power system reliability.

Studying the past blackouts can be helpful to prevent future ones [5–9]. But neither studying nor analyzing past events can prevent all future blackouts [10–15]. The aim is focused to reduce the probability of their occurrence and the extent of their consequences. This leads to improved power system reliability.

2.2 Selected Recent Blackouts: Causes and Consequences

The following recent blackouts are selected for a brief overview of the most important related facts investigating the causes and consequences:

- USA and Canada blackout, Aug 14, 2003
- Austria blackout, Aug 27, 2003
- London blackout, Aug 28, 2003
- Southern Sweden and Denmark blackout, Sep 23, 2003

M. Čepin, *Assessment of Power System Reliability*,
DOI: 10.1007/978-0-85729-688-7_2, © Springer-Verlag London Limited 2011

- Italian and Swiss blackout, Sep 28, 2003
- Greece blackout, July 12, 2004
- Moscow blackout, May 25, 2005
- Blackout, Nov 4, 2006

2.2.1 USA and Canada Blackout, Aug 14, 2003

USA and Canada blackout caused the outage where more than 60 GW of load was lost and 50 million people were affected [2, 13]. The loads in Northern Ohio were high because of use of air conditioning in an ordinary August afternoon. Import of 2,000 MW caused the system to consume high levels of reactive power. Two Cleveland power plants were already in shutdown. Therefore, the loss of the Eastlake 5 unit further depleted critical voltage support for the Cleveland area. Transmission line loadings were high, but not critical.

Stuart–Atlanta 345-kV line tripped offline because of tree contact. This line also had no major effect on electrical system. Midwest Independent System Operator state estimator was unable to assess system conditions and to perform contingency analyses of generation and line losses within its reliability limits. Consequently, it could not determine that with Eastlake 5 down, other transmission lines would overload, if a major transmission line would be lost.

Following the loss of Eastlake 5, the operators became concerned about voltage levels. They lost the alarm function and they remained unaware that their electrical system condition was beginning to fall. Three 345-kV lines failed with power flows at or below their emergency rating. Each failure was a result of contact between a line and a tree. As each line failed, loads on remaining lines increased. Voltages on the rest of system degraded further. An hour and a half after the alarm failed, the energy management system realized their situation. The operators did not know how much of their system was lost and how many line outages occurred after the trip of Eastlake 5, so they took no actions to return the system to reliable state.

The Midwest Independent System Operator realized that the system was endangered. The only way that the blackout might have been avoided, would have been to drop at least 1,500–2,500 MW of load around Cleveland and Akron. But it was not done.

The loss of 345-kV lines in northern Ohio caused its underlying network of 138-kV lines to fail, which was leading to a loss of 345-kV line Sammis–Star. Each of 345-kV lines in the Cleveland area tripped out. This increased loading and decreased voltage on the 138-kV system putting those lines into overload. Soon, the first of sixteen 138-kV lines failed shutting down customers in Akron area, dropping about 600 MW of load. The loss of the Sammis–Star line triggered the cascade.

Large electricity flows were moving across power system from generators in south to load centers in northern Ohio before the collapse. This pathway became unavailable. That is why power surged in from western Ohio, Indiana, and Pennsylvania through New York and Ontario around the northern side of Lake Erie. As lines were already overloaded, some of them began to trip. These power surges caused lines in neighboring areas to overload and to trip.

Then, the line trips progressed. Also lines in western Ontario became overloaded and tripped. The entire north eastern United States and the province of Ontario then became a large electrical island, separated from the rest of the Eastern Interconnection. This large island quickly became unstable, as there was not sufficient generation in operation.

Systems to the south and west of the split remained intact and were mostly unaffected by the outage. The cascade was isolated when the northeast was split from the eastern interconnection.

The large electrical island in the northeast was unstable. Many lines and generators tripped and the area was divided into several electrical islands. Some of those islands were unbalanced and some lines tripped. Some of the electric islands reached the balance and restored the loads.

More than 260 power plants had been lost, and tens of millions of people in the United States and Canada were left without electric energy. About 61,800 MW of load were lost. About 48,800 MW were restored in the next morning [2, 13]. The restoration varied between utilities from 29 h to 8 full days [2, 13].

The major blackout was caused by deficiencies in specific practices, equipment, and human decisions that coincided that afternoon. There were three major groups of causes:

- Unsuitable corporation awareness
- Corporation failed to manage tree growth in its transmission rights-of-way
- Failure of the interconnected grid reliability organizations to provide effective diagnostic support

2.2.2 Austria Blackout, Aug 27, 2003

On August 27, during a test procedure of safety valves, an improper setting of a switch led to an automatic switch-off of Krško Nuclear Power Plant in Slovenia.

Soon, a 400-kV breaker feeder tripped by protection. The load flow of Héviz lines increased to more than 1,300 A. Protection sent tripping command to all breakers in Tumbri station and to Héviz. The line Hëviz–Tumbri was disconnected. Then, the load flow in Austria increased up to 115% of thermal rating. Two lines at Bisamberg and one line at Dürnrohr tripped. The tripping of all lines with Hungary was prevented.

As a result of all failures, power flow started to flow from Czech Republic to Germany, instead of the scheduled flow from Germany to Czech Republic.

Flow increased from 1,100 to 1,500 MW and from 1,260 to 1,600 MW between Czech Republic and Slovakia [2].

Flow directions changed. The reactive power flow increased from Slovenia toward Austria. A large voltage drop was detected at Krško substation. Also the 110-kV network in Germany experienced a 108% overload because of increasing cross-border flows. This caused a network failure between Etzenricht and Pleinting. As a result, the power flew from Slovakia to Austria through Hungary.

The loss of Tumbri station increased the import from Hungary to Austria. The voltage drop in southern Austria was stabilized by additional power deliveries from all operational power plants in the region. Line Héviz-Tumbri was closed for 2 h and 6 min after the Krško Nuclear Power Plant trip. The three Austrian lines were closed for almost 3 h.

The occurrence of each single fault event would not have initiated such consequences. But the combination of high power flows, trip of a power plant Krško, and the welded contact of line protection in Tumbri caused a blackout.

2.2.3 London Blackout, Aug 28, 2003

In the evening of Aug 28, 2003, the southern part of London faced a power blackout. A 724 MW of supplies were lost, which is around 20% of total London supplies at that time. This affected around 410,000 customers with supplies being lost to parts of London underground and railway network [3].

The transmission system in London consists of four substations: Littlebrook, Hurst, New Cross, and Wimbledon. On that day, scheduled maintenance was performed on one circuit from Wimbledon to New Cross and on one from Littlebrook to Hurst. The maintenance had been planned.

The Electricity National Control Center received an indication that a transformer or shunt reactor at Hurst substation was in trouble and could trip. The gas had probably accumulated within the oil inside the transformer or shunt reactor, which could lead to equipment failure.

The Electricity National Control Center notified the alarm and disconnected the distribution system from the transformer. The disconnecting sequence began and it correctly disconnected Hurst substation from Littlebrook substation. This enabled a safe shutdown of the transformer and shunt reactor, but left Hurst to be supplied from Wimbledon through New Cross. A few seconds after the switching, automatic protection equipment on the circuit number two from Wimbledon to New Cross operated. An incorrectly installed automatic protection relay misread a change of power flow and automatically disconnected an area of south London from the rest of the network.

Fortunately, the London blackout was isolated to a single circuit and areas beyond that circuit have not lost power.

2.2.4 Southern Sweden and Denmark Blackout, Sep 23, 2003

On Sep 23, 2003, the Nordic power system was hit with the most severe disturbance in last 20 years. The southern Sweden and the eastern part of Denmark, including Copenhagen, were blacked out. Around 4 million people were affected [3].

Before the interruption, the average power use in Sweden was around 15 GW, which was quite usual because of warm weather. The nuclear energy in the affected area was limited because of annual maintenance and delayed restarts of some units, as nuclear safety was a prior significance. Only minor hydro and local generation was in service in southern Sweden. At that time, Danish grid was scheduled to an export of 400 MW to Sweden. Two 400-kV lines in the area were out of service because of maintenance work. Also high-voltage direct current lines to Poland and Germany were out.

The unit 3 in Oskarshamn Nuclear Power Plant started to pull back by manual control from its initial 1,175 MW to only 800 MW because of internal problems with feedwater system. The intention to solve problems failed and the reactor was shutdown.

Loss of 1,200-MW unit can be regarded as standard contingency. Reactive reserves shall be available to cope with this level of power loss without any other interruptions. After a normal transient in frequency and automatic activation of reserves from Norway hydro power, northern Sweden and Finland grid returned to stable conditions. Voltages in the southern part had dropped for around 5 kV, but remained uncritical. The frequency was automatically stabilized a bit below the normal operating limit.

A double busbar fault occurred in a 400-kV substation on western coast of Sweden. Two 900-MW units in nuclear power station Ringhals with 1,750 MW were disconnected.

The disconnection of those two units caused power oscillations in the power system, very low voltage and large overload. Soon, the oscillations faded out and the system seemed to be stabilized. But the demand in the area recovered from initial reduction following the voltage drop by action of many feeder transformer tap-charges. The voltage on the 400-kV grid collapsed. The grid was split up into two parts. The southern part remained interconnected but suffered from massive defect of generation. Soon, the frequency and voltage dropped to levels where generators and other grid protections reacted and this entire subsystem collapsed. Some small islands remained around some small hydro stations.

The loss of power supply was 4,500 MW in Sweden and 1,850 MW in Denmark. North of this area the power system was untouched. The hydro power in Norway, northern Sweden, and Finland was available to pick up the recovery of the demand. Emergency restoration proceeded. Lines and substations were energized to build up the grid from north to south. The National Grid Control Center managed to energize the 400-kV grid within less than 1 h. The restoration in Denmark had to be supported from Sweden. By that evening, almost all supplies in

Sweden and Denmark were resumed. The total non-supplied demand was around 10 GWh in Sweden and 8 GWh in Denmark.

The cause of this huge blackout was a very severe fault only a few minutes after a more ordinary but still significant fault.

2.2.5 Italian and Swiss Blackout, Sep 28, 2003

On Sep 28, 2003, a blackout affected more than 56 million people in Italy and Switzerland. In the early morning of that day, a cascading series of line trips led to the isolation and eventual blackout of Italian electricity distribution network. The cause of the failure occurred in the Swiss transmission system. A 380-kV line between Mettlen and Lavorgo was loaded at 86% of its capacity. As the temperature of line increased, the line slowly began to sag close to nearby trees. A flashover occurred and the Swiss could not close the line in sufficient time [6]. Because of the failure on Mettlen-Lavorgo line, a Sils-Sosa line became overloaded. As before, the temperature of this line also increased and the line sag to nearby trees, and the Swiss were unable to prevent another flashover. Load then increased on the 220-kV Airolo-Mettlen line, which was disconnected by automated protection devices. All these processes affected Italian power system.

As Italy is not able to produce enough electrical power for its domestic use, it must be imported from other countries, including Switzerland. As the Swiss line failed, it led to sudden loss of voltage in Italy. Italian network was placed out of synchronization with the rest of its partners in the UCTE. Automatic protection once again separated all remaining lines between the Italian grid and UCTE neighbors.

The Italian power system collapsed. The blackout lasted for more than 48 h. The mobile phone system began to fail, other areas of networks became overloaded and many UPS sources failed or ran out of battery power. Around 30,000 people were trapped on trains, many other hundred passengers were captured in underground train stations, and many flights in Italy were cancelled. The blackout also caused three deaths [6].

2.2.6 Greece Blackout, July 12, 2004

A blackout hit Athens and southern Greece on July 12, 2004. The load peak is expected during the noon hours of the hottest days.

The Hellenic system, especially during summer, is disposed to voltage instability. The problem is related to maximum power transfer from generating areas in the North and West Greece to the main load center in the Athens. The biggest problem is the electrical distance between generation and load, consisting of generator step-up transformers. The weakest part of the system has become the area from Central Greece to the north of Athens area.

As Olympic Games were being organized in Athens, in September 2004, upgrades in power grid were necessary. This included new 400/150-kV autotransformers, new capacitor banks in medium and high-voltage buses, and a new 150-kV connection between the major power station of Lavrio and Athens through the substation Argyroupoli. Many of the planned upgrades were not done by July 12, 2004. The two high-voltage autotransformers were not installed and several capacitors banks throughout Athens and Central Greece were also not in place [7].

In addition, many other transmission elements were unavailable on this day because of failures and repairs. The circuit of 150-kV line from Lavrion to Pallini was out of order, two 150-kV cables connecting AHSAG power station and the port of Athens were out, and so was one 150-kV circuit between Koumoundouru substation and the power station of AHSAG.

The unit 2 (rated 300 MW) of the Lavrio power station in the Athens area was lost because of failure. The failure was repaired and later the power station was put back to work. During that time, voltages in Athens area were dropping and were reaching 90% of the nominal value. As soon as Lavrio-2 was synchronized, the voltage decline stopped. However, when Lavrio-2 was still in process of achieving its technical minimum and on manual control it was lost again. The new loss of Lavrio-2 brought power system to emergency state, as the other generating stations in the area tried to keep up with the demand.

The load shedding of 100 MW was requested by Hellenic Transmission System Operator Control Center. A disconnection of 80 MW was achieved manually. This was not enough to stop the voltage decline, so further shedding action of 200 MW was requested, but it was unsuccessful. The unit 3 of Alivieri power station tripped and a minute later the remaining unit in Alivieri was tripped manually. After all the voltage collapsed, the system was split. All the areas of Athens and Peloponnese were disconnected, which lead to the blackout. The restoration of the Athens area was fast in only 2–3 h.

The blackout knocked-out air conditioning during extremely hot temperatures. The outage also caused many traffic jams and accidents as all the traffic lights were out of order. Many passengers were forced to leave subway and walk, and the fire department received many calls about people being trapped in elevators. The blackout also caused some cell phone networks to overload.

2.2.7 *Moscow Blackout, May 25, 2005*

The blackout on May 25 affected nearly 4 million people who got stuck in elevators, on subway trains, on subway stations, and in street traffic. Economic damage was estimated to 70 million American dollars.

The cascade power stations on the Volga and Kama rivers are one of the largest renewable energy sources in southwest Russia. However, the equipment is more than 40 years old and completely exhausted [10].

After decades of proper working, two electrical transformers in the Chagino power substation suddenly tripped. Chagino substation lost its transformers, switches, and insulation because of equipment damage. The entire substation Chagino component broke down. Moscow power circle of 500 kV broke up because of the power loss in the high-voltage liner during the night from May 24–25, 2005.

A few high-voltage lines of 110 and 220 kV were overloaded and many automatic power breaks in high-voltage lines occurred. Still some of 100-kV lines remained in operation, but were still overloaded, which lead to network collapse. The voltage level in the Southern part of Moscow power system suffered voltage drop. As the operational personnel response was not quite adequate, a blackout occurred.

Russians were shocked that their country, so rich in energy sources, did not have a proper power infrastructure to keep up with their demand.

2.2.8 *Blackout, Nov 4, 2006*

An incident on the continental European electricity network on Nov 4, 2006 has led to blackouts all over the grid [8, 9]. Several failures occurred leading to power blackout that affected 10 million people in Germany, France, Belgium, Spain, and Austria. Three main reasons were revealed as the causes for the event.

- The German electricity transmission system operator, who was origin of the fault, did not have the security procedure in place.
- Other European transmission system operators did not receive information on the actions taken by the German transmission operator.
- The lack of reliability level and the operation of the grid are also to be blamed.

The fault originated from Northern Germany, from the control area of German electricity transmission system operator. There were four transmission system operators in Germany including Eon Netz, RWE Transporterz Strom GmbH, Vattenfall Europe Transmission GmbH, and EnBW Transportnetze AG. Organizational changes took place in recent years.

A high-voltage line Conneforde–Diele had been switched off to let a ship pass underneath. The disconnections of lines across the river Ems happen regularly, which allows large ships to safely cross the river without disconnecting electrical lines. The double-circuit 380-kV line is usually disconnected for between two and four hours. Eon Netz approved the request by the shipyard on October 27, after they had previously performed analysis of the impact of switching off the 380-kV line. The results confirmed that the grid would be highly loaded, but secure. Furthermore, Eon Netz informed TenneT and RWE about the provisional agreement. As a result of coordination, all transmission system operators agreed to reduce the cross-border transmission capacity from Eon Netz to TenneT.

On November 4, TenneT decided to further reduce the capacity between Germany and the Netherlands because of the wind forecast. But the reduction was made only on the capacity from RWE. Also on November 3, the shipyard requested Eon Netz to do the disconnection of the line 3 h earlier. Eon Netz agreed, but no analysis or simulations were carried out. At this time, RWE and TenneT were not notified of the advanced timing. RWE and TenneT were notified next day. At that time, it was no longer possible to reduce the cross-border exchange program. So TenneT agreed with Eon Netz and RWE to change the tap position on the phase shifter in Meeden to reduce expected high flows. Because of construction work on the 380-kV Borken substation, the substation operated in a two-busbar mode. Usually it operates in only one. That meant that power flows were not possible from East to West in this region.

Eon Netz switched off the first circuit of the 380-kV line across river Ems, only a minute after Eon Netz received several warning messages about the high-power flow on the lines Elsen-Twistetal and Elsen-Bechterdissen. One warning message indicated that the power limits were about to be reached, but Eon Netz took no immediate action.

The load on the 380-kV line Landesbergen-Wahrendorf increased by 100 MW and the current to 160 A. Soon, the power increased to 1,900 MW, thereby exceeding the warning value of 1,800 A. As no proper actions were made, the automatic protection device disconnected the line. This resulted in overloading and tripping of the 220-kV line Bielefeld–Güterlosh, and later the 380-kV line Bechterdissen–Elsen. Other lines from North to South across Germany, Austria, Croatia, Hungary, and other countries tripped within the next few seconds.

As a result of all these failure, the European transmission grid was split into three zones. The north-east area had an over-frequency; in the western area, there was an under-frequency; and south-eastern area also suffered a frequency drop. In these areas, consumers were disconnected from the network. In the evening, the power supply was restored for consumers.

2.3 Blackout Prevention

The relative time of action for different types of events, from normal to extreme, varies depending on type and speed of the failure and the need for coordination. Recent blackouts show the need for investment and deployment of well-defined and coordinated defense plans. Any investment should consider the long-term impact connected to system adjustments and the supported system studies.

There is no solution to prevent blackouts, but many things can be made to minimize disturbances including: analyses and audits, preventive and corrective actions, improved monitoring, diagnostics and control center performance, secure real-time operating limits on a daily basis, study protection coordination and design, test relays, relay applications and system protection schemes, study dynamic voltage and transient stability, assess condition of aging infrastructure and improve maintenance procedures, and secure control and protection system and reinforced network.

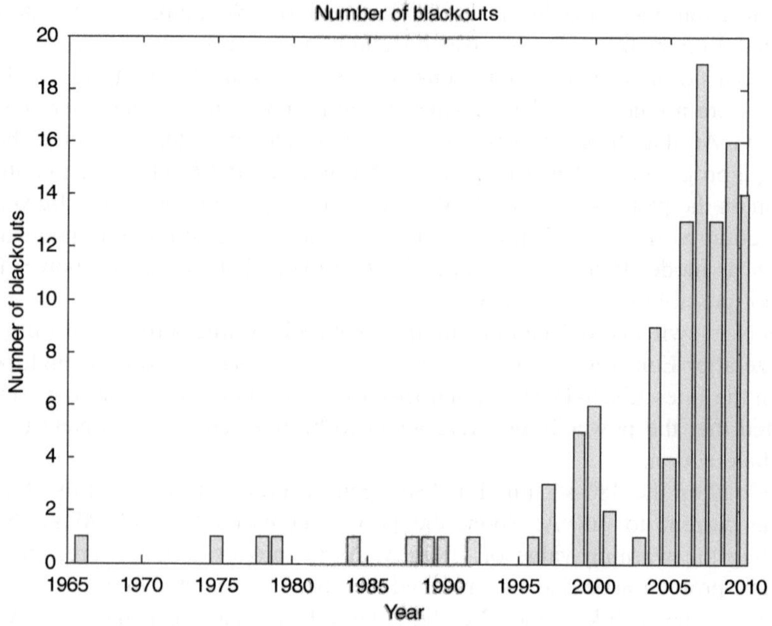

Fig. 2.1 Number of blackouts in the last decades

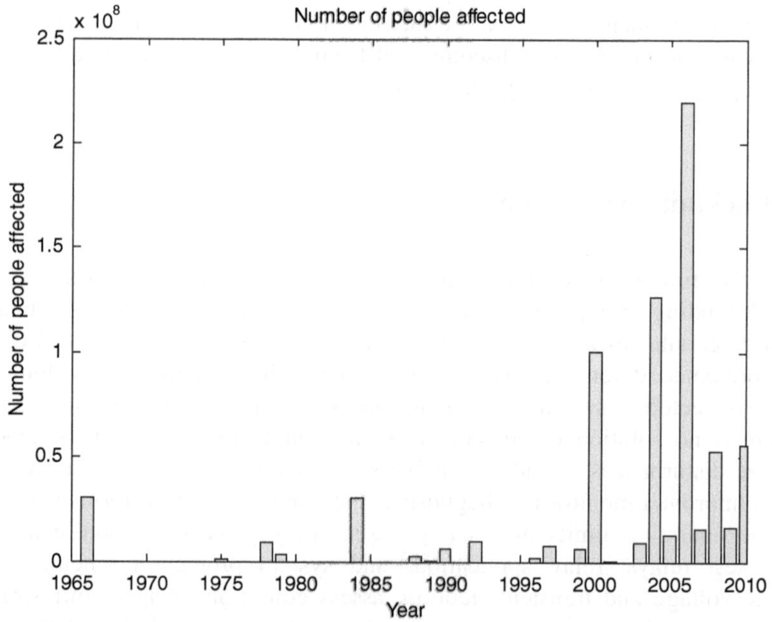

Fig. 2.2 Number of people affected because of blackouts in the last decades

2.4 Blackout Consequences

The probability of future blackouts is very low and difficult to predict. Their impact can be destructible. Estimation of the future risks is an important task, not only to prevent huge economic damage but also to protect human lives. Figure 2.1 shows the number of blackouts in the last decades. The highest number of power blackouts occurred in 2006.

Figure 2.2 shows the number of people affected because of blackouts, which seems to be increasing. Based on Figs. 2.1 and 2.2, one can conclude that the consequences of blackouts may be larger and larger. A large number of activities in the reliability of power systems should take place in order that the probabilities and consequences of blackouts are limited.

References

1. Holmgren ÅJ (2005) Electricity case: risk analysis of infrastructure systems—different approaches for risk analysis of electric power systems, CREATE Report, New York University-Wagner Graduate School, Institute for Civil Infrastructure Systems
2. Eurelectric (2004) Power outages in 2003: task force power outages (ref 2004-181-0007). Union of the Electricity Industry. Brussels
3. World Energy Council (2003) Focus: blackouts. World Energy Council, London. http://www.worldenergy.org/focus/blackouts/390.asp. Accessed 27 July 2010
4. Leahy E, Tol RSL (2011) An estimate of the value of lost load for Ireland. Energ Policy doi: 10.1016/j.enpol.2010.12.025
5. Larsson S, Ek E (2004) The blackout in southern Sweden and eastern Denmark. In: Proceedings of the IEEE PES general meeting, Denver, CO
6. Johnson CW (2007) Analyzing the causes of the Italian and Swiss blackout, 28 Sep 2003. In: Proceedings of 12th Australian conference on safety critical systems and software conference, Australian Computer Society
7. Vournas C (2004) Technical summary on the Athens and Southern Greece blackout of 12 July 2004. National Technical University of Athens
8. Li C, Sun Y, Chen X (2007) Analysis of blackout in Europe on 4 Nov 2006. In: Proceedings of the 8th international power engineering conference, IPEC, RPS, pp 939–944
9. Federal Network Agency for Electricity, Gas, Telecommunications, Post and Railways (2007) Report on the disturbance in the German and European power system on the 4 Nov 2006. Bundesnetzagentur, Bonn
10. Chubais A (2005) Post-transition Russia: risks and rewards. In: Proceedings of the IX annual equity conference, Moscow, Russia
11. Novosel D, Begovic MM, Madani V (2004) Shedding light on blackouts. IEEE Power Energ Mag 2(1):32–43. doi:10.1109/MPAE.2004.1263414
12. Periera L (2004) Cascade to black. IEEE Power Energ Mag 2(3):54–57
13. US–Canada Power System Outage Task Force (2004) Final report on the 14 Aug 2003 blackout in the United States and Canada: causes and recommendations
14. Swiss Federal Office of Energy (2003) Report on the blackout in Italy on 18 Sep 2003. SFOE
15. Casazza J, Delea F (2003) Understanding electric power systems: an overview of the technology and the marketplace. Wiley, Hoboken, NJ

Chapter 3
Definition of Reliability and Risk

There are two times in a man's life when he should not speculate: when he can't afford it, and when he can
Mark Twain

3.1 Introduction About Terminology

Component is a piece of equipment, which is not further divided for the analysis purposes. *System* is a group of components, which are associated or connected to perform a specific function or more functions. *Equipment* is a term covering both components and systems, which depends on particular consideration. An item may be any part, subsystem, system, or equipment that can be individually considered and separately tested. The term item can correspond to the term component, but the term item may be of wider meaning as the term component always means something physical, whereas the item can represent also something that is not necessarily physical, e.g., procedure step in operating procedures.

Sometimes, such terminology is incomplete, because it is difficult to treat the human actions as components of the system or the items of the system, but they are actually contributing in some similar way to the system reliability as physical components. The terms connected with human actions, such as operating procedures or testing procedures or emergency procedures, are part of system documentation for certain systems and human errors, i.e., failures of human actions can be a source of contribution to reliability calculations. On one side, it is difficult to say that human error is a component or an item, but on the other side, the human errors contribute to systems reliability considerably and therefore they should not be avoided.

A failure is any inability of a part or equipment to carry out its specified function. A fault is an event of wider meaning than the failure and includes the failures of specific equipment and their related features needed for the proper operation of the equipment. The difference between the failure and the fault can be explained on the following examples.

If the relay opens as required when a current runs or stops running through a circuit, this is a success. If the relay fails to open under these circumstances, this is a relay failure. If the relay opens at the wrong time because of the improper functioning of some upstream component, it is not a relay failure, but relay may well cause the entire circuit to enter into an unsatisfactory state: a fault.

M. Čepin, *Assessment of Power System Reliability*,
DOI: 10.1007/978-0-85729-688-7_3, © Springer-Verlag London Limited 2011

If the signal for a switch closure does not arrive timely to the switch control circuit, one cannot consider this a failure of a switch, although it is a fault of the switch that has not closed.

3.2 Reliability and Availability

Because of the many different operational requirements and varying environments, reliability means different things to different people. The generally accepted definition of reliability defines the reliability as the characteristic of an item expressed by the probability that it will perform a required function under stated conditions for a stated period of time [1].

$$R(t) = P\Big[E \text{ did not fail in time interval } [0,t]\Big] \tag{3.1}$$

Unreliability is a complement of reliability.

The generally accepted definition of availability defines the availability as the characteristic of an item expressed by the probability that it will perform a required function under stated conditions in a stated moment of time [1]. Mostly, the time intervals when the equipment is available and the time intervals when the equipment is unavailable are considered.

$$A(t) = P\Big[E \text{ did not fail in time } t\Big] \tag{3.2}$$

Unavailability is a complement of availability.

The mathematical definition of reliability is related to the probability density function $f(t)$, which is presented in Chap. 4 on probability theory.

Reliability $R(t)$ is the probability that the variable is at least as large as t. For continuous random variable, the related equation is the following:

$$R(t) = \int_{t}^{\infty} f(t)dt \tag{3.3}$$

For discrete random variable, the related equation is the following:

$$R(t_i) = \sum_{i=1}^{i=k} f(t_i) \tag{3.4}$$

The term reliability is divided into two terms when dealing with the power systems [2–5]. Those two terms are adequacy and security. The adequacy is related to the existence of sufficient generation of the electric power system to satisfy the consumer demand. The security is related to the ability of the electric power system to respond to transients and disturbances that occur in the system.

3.3 Risk

Risk is a combination of a probability for an accident occurrence and resulting negative consequences. Risk is often reserved for random events with negative consequences to human life and environment [6–8].

The mathematical definition of risk is represented by equation:

$$\text{Risk} = C \times P \tag{3.5}$$

where C is the extent of consequences and P is the probability of the considered event.

The extent of consequences can represent the number of death of people in a determined time interval. Or this can be a number of injuries in a time interval or the amount of money or property or other resources lost in the time interval.

The risk can be presented on a figure. Figure 3.1 shows the risk as a measure of probability of an event and its consequences. The area of small risk is the area where the probability is small and the consequences are small. Several attempts to distinguish acceptable risk from unacceptable risk with a quantitative risk limit are shown.

The risk criterion is a term that may distinguish between what is considered as an acceptable level of safety and what it is not [9–25]. Development of the risk criteria is one of prerequisites for risk-informed decision making. The methods and tools are advanced in nuclear field and air and space industry, but they are more and more applied also in other areas.

Because it is difficult to numerically distinguish the acceptable from unacceptable risk with a straight quantitative measure, the approach as low as reasonably practicable (ALARP) has been developed, which is similar to the approach as low as reasonably achievable (ALARA). Figure 3.2 shows the approach that introduces new area of tolerable risk, where the risk decrease is desired, but is not required if the related costs highly exceed the benefits of decreased risk. Figure 3.3 shows the approach ALARP.

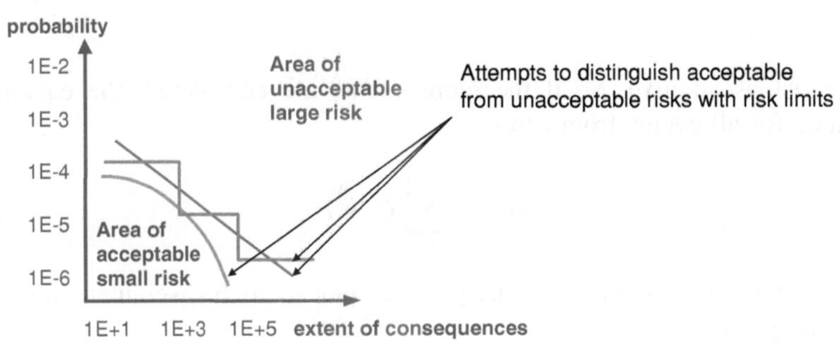

Fig. 3.1 Risk representation

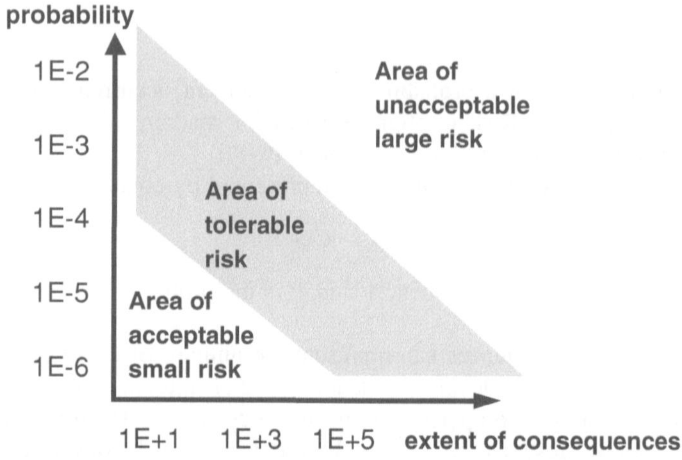

Fig. 3.2 Tolerable risk region

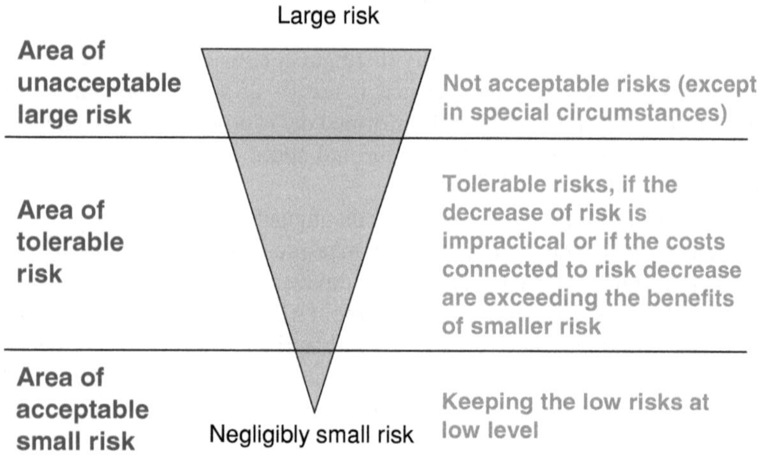

Fig. 3.3 Approach as low as reasonably practicable (ALARP)

The risk is additive. So if the more events are considered, the equation is expanded for all events from i to ii.

$$\text{Risk} = \sum_{i=i}^{i=ii} C_i \times P_i \tag{3.6}$$

Some definitions require the large consequence to be weighted more with additional power of n.

$$\text{Risk} = C^n \times P \tag{3.7}$$

The most important general simplification no matter, if the consequences are weighted with additional power of n or not, says that in the case of constant consequences, the probabilities of events represent the measure of risk.

3.4 N−1 Reliability Criteria

The $n-1$ reliability criterion is used in transmission system planning [9, 10]. The $n-1$ criterion requires that the loss of any single element in the power system should not prevent the supply of the electric power. The acceptable system conditions shall exist after the loss of a major system equipment such as a generating unit, transmission line, or transformer. This is required even if the line with the highest capacity is the one that goes out of service. This is required even if the largest generating unit is lost. Mostly, the loss of the largest generating unit is a limiting event for which it is required to proof that the power system is able to withstand it even if it occurs at the worst possible moment of time. Normal operation must be restored within minutes after a fault. Sufficient fast and slow operational reserves must be available.

Certain analogy exists with single-failure criterion, a design criterion known in the field of nuclear safety. Single-failure criterion requires such design of any safety system that no single failure of any equipment can jeopardize realization of safety functions [11, 12].

References

1. Villemeur A (1992) Reliability, availability, maintainability and safety assessment, methods and techniques. Wiley, New York
2. Billinton R, Allan R (1996) Reliability evaluation of power systems. Plenum, New York
3. Brown RE (2002) Electric power distribution reliability. Marcel Dekker, New York
4. Short TA (2006) Distribution reliability and power quality. Taylor & Francis, Boca Raton
5. Dummer GWA, Tooley MH, Winton RC (1997) An elementary guide to reliability. Butterworth-Heinemann, Oxford
6. Berg HP, Gortz R, Schimetschka E (2003) Quantitative probabilistic safety criteria for licensing and operation of nuclear plants. BFS-SK-03/03, BFS
7. Jehee J, deWitt H, Patrik M, Bareith A, Čepin M et al (2002) Risk goals and system targets. Project report Ch. 2.3, IJS
8. Gortz R (2001) Risk targets and reliability goals. Lecture 16, EUROCOURSE PSARID 2001, GRS
9. Reppen DN (2004) Increasing utilization of the transmission grid requires new reliability criteria and comprehensive reliability assessment. In: Proceedings of the eighth international conference on probabilistic methods applied to power systems (PMAPS 2004), Iowa State University, Ames, New York, pp 933–938

10. Chowdhury AA, Koval DO (2006) Probabilistic assessment of transmission system reliability performance. IEEE
11. IEEE (1994) IEEE standard application of the single-failure criterion to nuclear power generating station safety systems
12. ANSI/IEEE Std 379 (1988) Standard application of the single-failure criterion to nuclear power generating station safety systems. ANSI/IEEE
13. Holmberg J, Puikkinen U, Rosquist T, Simola K (2001) Decision criteria in PSA applications. NKS-44
14. YVL-28 (2003) Probabilistic safety analysis in safety management of nuclear power plants. STUK
15. GS-1.14 (2002) Criteria for the performance of probabilistic safety assessment applications
16. Čepin M (2007) The risk criteria for assessment of temporary changes in a nuclear power plant. Risk Anal 27(4):991–998
17. Health and Safety Executive (1992) Safety assessment principles for nuclear plants. London
18. Čepin M, Jordan Cizelj R, Mavko B (2004) Qualitative and quantitative criteria for application of probabilistic safety assessment in decision-making (in Slovenian), IJS-DP-8861. Institute Jožef Stefan
19. IAEA-TECDOC-1436 (2005) Risk informed regulation of nuclear facilities: overview of the current status. IAEA, Vienna
20. NEA/CSNI/R(2002)18 (2002) The use and development of probabilistic safety assessment in NEA member countries. NEA
21. ASME RA-S-2002 (2002) Standard for probabilistic risk assessment for nuclear power plant applications. ASME
22. TR-105396 (1995) PSA applications guide. Electric Power Research Institute
23. Samanta P, Kim IS, Mankamo T, Vesely WE (1995) Handbook of methods for risk-based analyses of technical specifications. NUREG/CR-6141, NRC
24. Vesely W, Dugan J, Fragola J et al (2002) Fault tree handbook with aerospace applications. National Aeronautics and Space Administration
25. Kumamoto H, Henley EJ (1996) Probabilistic risk assessment and management for engineers and scientists. IEEE, New York

Chapter 4
Probability Theory

The probable is what usually happens

Aristotle

4.1 Introduction

Probability theory is a part of mathematics that aims to provide insights into phenomena that depend on chance or on uncertainty [1–8]. The mathematical theory of probability is very sophisticated. A basic treatment of probability from a perspective of engineers who are going to use the probability theory as a support for the practical reliability analyses is presented.

Probability theory provides an analytical treatment of events. The first approach is to define probability in terms of frequency of occurrence, as a percentage of successes in a large number of similar situations. This approach is suitable where a large number of similar events are expected to occur. For example, in many tosses of a two-sided coin, the head appears at the top in 50% of tosses, or the chance of tossing a head of a coin is one out of two or 50%. For example, what is the probability of drawing a blue marble or a red marble out of a jar containing three blue, one red, and six green marbles assuming that the marbles are all the same size and equally weighted and they are well mixed so that every marble is equally likely to be picked? Picking a red or a blue marble ($4 = 3 + 1$) out of a jar of ($10 = 3 + 1 + 6$) marbles gives a success probability of 4 out of 10 or 0.4.

The second approach is subjective approach. If there are not many similar events, or if there is only one time event in question, then the probability of the event may express the subjective belief about the event. For example, one can be 80% sure that his or her car has been technically acceptable, or he or she is 90% sure that he or she has a coin in his or her pocket.

The mathematical representation of probability theory starts with set theory and basic probability concepts.

M. Čepin, *Assessment of Power System Reliability*,
DOI: 10.1007/978-0-85729-688-7_4,

4.2 Set Theory

The basis of probability theory lays in defining the triplet of the experiment, the outcome of the experiment, and the sample space of the experiment [3, 4].

An experiment is a process, for which the result is uncertain, such as tossing a coin, for example. An outcome is the result of one execution of the experiment, which can be tossing a head or tossing a tail, if the experiment is tossing a coin. Because of uncertainty associated with the process, repetitions or trials of a defined experiment would not be expected to produce the same outcomes. The set of all possible outcomes of an experiment is defined as the sample space. Sample spaces can contain discrete values (such as head or tail, if the experiment is tossing a coin) or values in a continuum (such as measurement of time between failures). An event E is a specified set of possible outcomes in a sample space S.

A set is a collection of the elements. If S is a set and $E1$ is an element of S, we write $E1 \in S$. If element $E1$ is not an element of S, we write $E1 \notin S$. A set can have no elements, in which case it is called the empty set, denoted by $\emptyset$. Sets can be specified in a variety of ways. If set S contains a finite number of elements, say $E1, E2, \ldots, En$, we write it as a list of the elements, in braces:

$$S = \{E1,\ E2,\ \ldots, En\}$$

If set S contains infinitely many elements, which can be enumerated in a list we write:

$$S = \{E1, E2,\ \ldots\}$$

and we say that S is infinite. For example, the set of even integers can be written as:

$$S = \{0,\ 2, -2,\ 4, -4,\ \ldots\}$$

and is infinite.

Alternatively, we can consider the set of all elements E that have a certain property P, and write it as

$$S = \{E;\ E \text{ satisfies } P\}$$

The expression means the set of all elements E, such that E satisfies property P. For example, the set of odd integers can be written as

$$S = \{k;\ (k+1)/2 \text{ is integer}\}$$

The set of all scalars x in the interval [0, 1] can be written as

$$\{x;\ 0 \le x \le 1\}$$

Note that the elements x of this set take a continuous range of values, and cannot be written down in a list; such a set is said to be uncountable.

If every element of a set S is also an element of a set T, we say that S is a subset of T, and we write $S \subset T$. If $S \subset T$ and $T \subset S$, the two sets are equal, $S = T$. A universal set is denoted by Ω and contains all elements that could

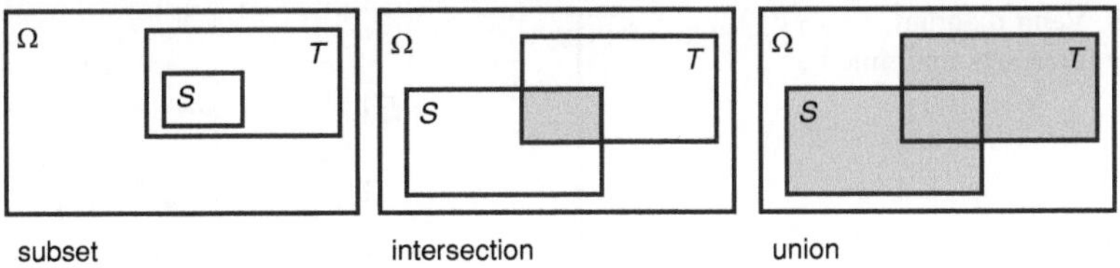

Fig. 4.1 Venn diagram showing subset, intersection, and union

conceivably be of interest in a particular context. Having specified the context in terms of a universal set Ω, we only consider sets *S* that are subsets of Ω.

Most events of interest in practical situations are compound events, formed by a composition of two or more events. Composition of events can occur through the complement of events or through union of events or through intersection of events or through some combination of these. For two events, *E*1 and *E*2, in a sample space *S*, the union of *E*1 and *E*2 is defined to be the event containing all sample points in *E*1 or *E*2 or both, and is denoted by the symbol ($E1 \cup E2$). Thus, a union is simply the event that either *E*1 or *E*2 occurs.

In other words, the union of two sets *S* and *T* is the set of all elements that belong to *S* or *T* (or both), and is denoted by $S \cup T$.

Thus, $S \cup T = \{x;\, x \in S \quad \text{or} \quad x \in T\}$.

For two events, *E*1 and *E*2, in a sample space *S*, the intersection of *E*1 and *E*2 is defined to be the event containing all sample points that are in both *E*1 and *E*2, denoted by the symbol ($E1 \cap E2$). The intersection is the event that both *E*1 and *E*2 occur.

In other words, the intersection of two sets *S* and *T* is the set of all elements that belong to both *S* and *T*, and is denoted by $S \cap T$.

Thus, $S \cap T = \{x;\, x \in S \quad \text{and} \quad x \in T\}$.

Figure 4.1 shows three cases. The *left part* shows that the set *S* is a subset of set *T*, or written mathematically: $S \subset T$. The *middle part* shows shaded intersection of sets *S* and *T*, or written mathematically: $S \cap T$ is shaded. The *right part* shows shaded union of sets *S* and *T*, or written mathematically: $S \cup T$ is shaded.

Figure 4.2 shows a Venn diagram of nine outcomes and three events. The event *E*1 contains six outcomes, event *E*2 contains three outcomes, their union contains seven outcomes, and their intersection contains two outcomes. The event *E*3 is empty. It cannot occur.

The complement of an event *E* is the collection of all sample points in *S* and not in *E*. The complement of *E* is denoted by the symbol E' because of easier writing (or in some literature by E^* or in some literature by E^C) and is the outcomes in *S* that are not in *E* occur. Figure 4.2 shows that the complement of *E*1 is an event containing three outcomes.

The empty set or null set is a set containing no outcomes. Figure 4.2 shows that the event *E*3 is an empty set. The two events, *E*1 and *E*3, are said to be mutually exclusive if the event ($E1 \cap E3$) contains no outcomes in the sample space *S*. That is, the intersection of the two events is the null set. Mutually exclusive events are

Fig. 4.2 Venn diagram showing three sets and nine events

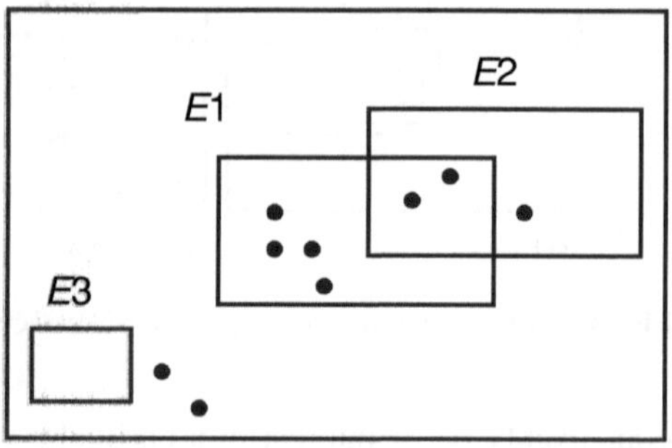

also referred to as disjoint events. Three or more events are called mutually exclusive, or disjoint, if each pair of events is mutually exclusive. In other words, no two events can happen together.

Set operations have several properties, which are elementary consequences of the definitions. Some examples are:

$$S \cup T = T \cup S \tag{4.1}$$

$$S \cup (T \cup U) = (S \cup T) \cup U \tag{4.2}$$

$$S \cap (T \cup U) = (S \cap T) \cup (S \cap U) \tag{4.3}$$

$$S \cup (T \cap U) = (S \cup T) \cap (S \cup U) \tag{4.4}$$

$$(S')' = S \tag{4.5}$$

$$S \cap S' = \emptyset \tag{4.6}$$

$$S \cup \Omega = \Omega \tag{4.7}$$

$$S \cap \Omega = S \tag{4.8}$$

Two particularly useful properties are given by De Morgan laws:

$$(X \cap Y)' = X' \cup Y' \tag{4.9}$$

$$(X \cup Y)' = X' \cap Y' \tag{4.10}$$

4.3 Basic Probability Concepts

A probabilistic model is a mathematical description of an uncertain situation. Two elements of a probabilistic model are:

- The sample space Ω, which is the set of all possible outcomes of an experiment. An event is a subset of the sample space.
- The probability law, which assigns to an event, which is a set A of possible outcomes, a nonnegative number called probability of A or $P(A)$ that encodes our knowledge or belief about the collective likelihood of the elements of A.

Each of the outcomes in a sample space has a probability associated with it. Probabilities of outcomes are seldom known; they are usually estimated from relative frequencies with which the outcomes occur when the experiment is repeated many times. Once determined, the probabilities must satisfy three requirements:

- The probability of each outcome must be a number between 0 and 1.
- If $A1$, $A2$, and $A3$ are disjoint events (their intersection is empty set), then the probability of their union is a sum of probabilities of those events: $P(A1 \cup A2 \cup A3) = P(A1) + P(A2) + P(A3)$. The rule is generalized also for larger number of events or for infinite number of events.
- The sum of probabilities of all outcomes in a given sample space is 1.

Example 1 Let us consider an experiment involving a single coin toss. There are two possible outcomes, tails (T) and heads (H). The sample space is $\Omega = \{H, T\}$, and the events are $\{H, T\}$, $\{H\}$, $\{T\}$, and $\emptyset$. If the coin is fair, which means that the event of getting heads and the event of getting tails are equally likely, we assign equal probabilities to the two possible outcomes and specify that probability of heads equals 0.5 and probability of tails equals 0.5. $P(\{H\}) = P(\{T\}) = 0.5$. The sum of probabilities of all outcomes in a given sample space is 1:

$$P(\{H,T\}) = P(\{H\}) + P(\{T\}) = 0.5 + 0.5 = 1$$

Example 2 Let us consider an event of tossing one square six-sided die. The sample space is six outcomes. $\Omega = \{1, 2, 3, 4, 5, 6\}$. The events of interest are: $T1 = \{1\}$, $T2 = \{2\}$, $T3 = \{3\}$, $T4 = \{4\}$, $T5 = \{5\}$, and $T6 = \{6\}$. If the six-sided die is square and made of homogeneous material, the probability of the event $T1$, which is the probability of outcome of tossing 1 is equal to 1/6. The same applies for all other events. The sum of the probabilities equals 1.

Example 3 Let us consider an event of tossing two coins. The sample space is four outcomes. Each outcome includes head (H) or tail (T) from each coin. $\Omega = \{HH, HT, TH, TT\}$. The events of interest are: tossing two heads: $A = \{HH\}$, tossing two tails: $B = \{TT\}$, tossing one heads, and one tails: $C = \{HT, TH\}$. The probability of outcome of tossing two tails equals to probability of tossing two heads and equals to 1/4, whereas the probability of outcome of one head and one tail is two times larger and it equals 1/2. The sum of probabilities of all outcomes equals 1.

The sample space of an experiment may consist of a finite or of an infinite number of possible outcomes. Finite sample spaces are simpler.

The probability of an impossible event (the empty or null set) is zero.

If the sample space consists of n possible outcomes that are equally likely (i.e., all single-element events have the same probability), then the probability of any event A is given by:

$P(A)$ = number of elements of A/n

In other words, the probability of an event is the ratio between the number of successful simulation experiments and the number of all simulation experiments.

The probability of the complement of E is given by: $P(E') = 1 - P(E)$.

In general, the probability of the union of any two events $E1$ and $E2$ is given by:

$$P(E1 \cup E2) = P(E1) + P(E2) - P(E1 \cap E2) \tag{4.11}$$

If the events $E1$ and $E2$ are mutually exclusive then $P(E1 \cap E2) = 0$ and $P(E1 \cup E2) = P(E1) + P(E2)$.

The probability of the intersection of two independent events $E1$ and $E2$ is given by:

$$P(E1 \cap E2) = P(E1) \times P(E2) \tag{4.12}$$

If both events are not independent, the conditional probability rules are to be followed.

If there are more events than two, the probability of union becomes more complex. The expression for the probability of union of three events $E1$, $E2$, and $E3$ is the following:

$$\begin{aligned} P(E1 \cup E2 \cup E3) = P(E1) + P(E2) + P(E2) - P(E1 \cap E2) - P(E1 \cap E3) \\ - P(E2 \cap E3) + P(E1 \cap E2 \cap E3) \end{aligned} \tag{4.13}$$

The general expression for determining the probability of union of more events is the following [9, 10]:

$$\begin{aligned} P\left(\bigcup_{i=1}^{n} E_i\right) = \sum_{i=1}^{n} P(E_i) - \sum_{i \le i_1 < i_2 \le n} P(E_{i_1} \cap E_{i_2}) + \sum_{i \le i_1 < i_2 < i_3 \le n} P(E_{i_1} \cap E_{i_2} \cap E_{i_3}) \\ - + \cdots + (-1)^{n+1} P\left(\bigcap_{i=1}^{n} E_i\right) \end{aligned} \tag{4.14}$$

4.4 Theory of Combinations

The definition of factorial and the Pascal triangle represent the background for the theory of combinations [11].

The continuous product of the first n natural numbers is called factorial n and is denoted by $n!$, which represents equation: $n! = 1 \times 2 \times 3 \times \cdots \times (n-1) \times n$. For $n = 0$, the factorial equals $1 : 0! = 1$.

	rows
1	n=0
1 1	n=1
1 2 1	n=2
1 3 3 1	n=3
1 4 6 4 1	n=4
1 5 10 10 5 1	n=5
1 6 15 20 15 6 1	n=6

Fig. 4.3 Pascal triangle

The Pascal triangle is a number triangle with numbers arranged in staggered rows such that the numbers in it represent binomial coefficients [11]. Figure 4.3 shows the Pascal triangle.

$$a_{n,r} = \binom{n}{r} = \frac{n!}{(n-r)!r!} \tag{4.15}$$

The Pascal formula shows that each subsequent row is obtained by adding the two entries diagonally above.

$$\binom{n}{r} = \frac{n!}{(n-r)!r!} = \binom{n-1}{r} + \binom{n-1}{r-1} \tag{4.16}$$

The rows contain the binomial coefficients. The sum of the elements of the ith row is 2^i. The shallow diagonals of Pascal triangle sum to Fibonacci numbers (1, 1, 2, 3, 5, 8, 13, . . .).

The first diagonal is of values of 1. The second diagonal has the counting numbers (1, 2, 3, . . .). The third diagonal has the triangular numbers (1, 3, 6, 10, 15, . . .). The fourth diagonal has the tetrahedral numbers (1, 4, 10, 20, . . .).

The theory of combinations determines the possible grouping of objects. There are three processes of interest: (i) permutations, (ii) combination, and (iii) variations, which are actually a union of the other two. One could say a permutation is an ordered combination.

4.4.1 Permutations

When the objects of a group are arranged in a certain order, the arrangement is called a permutation. In a permutation, the order of the objects is very important.

A permutation is a linear permutation if the objects are arranged in a line. A linear permutation is mostly called as a permutation.

If repetition is not allowed, the number of possibilities (N) is calculated considering a number of objects (n) and considering a number of objects in a set (r) according to equation:

$$N = \frac{n!}{(n-r)!} \quad (4.17)$$

In other words, the n is a number of things to choose from, and you choose r of them.

Example Calculation of a number of possible passwords. Let us imagine the alphabet with 26 letters and 10 possible digits (other characters are not considered), which means that 36 characters are possible for each of characters of the password ($n = 36$).

If a password is of one character ($r = 1$) and we choose from 36 characters ($n = 36$), the number of possible passwords is $N = 36!/(36 - 1)! = 36$. If a password is of two characters ($r = 2$) and we choose from 36 characters ($n = 36$), the number of possible passwords is $N = 36!/(36 - 2)! = 36 \times 35 = 1{,}260$. If a password is of six characters ($r = 6$) and we choose from 36 characters ($n = 36$), the number of possible passwords is:

$$\mathrm{N} = 36!/(36-6)! = 36 \times 35 \times 34 \times 33 \times 32 \times 31 = 1{,}402{,}410{,}240 \quad (4.18)$$

If repetition is allowed, the number of possibilities (N) is calculated considering a number of objects (n) and considering a number of objects in a set (r) according to equation:

$$N = n^r \quad (4.19)$$

Example Calculation of a number of possible passwords. Let us imagine the alphabet with 26 letters and 10 possible digits (other characters are not considered), which means that 36 characters are possible for each of characters of the password ($n = 36$).

If a password is of one character ($r = 1$) and we choose from 36 characters ($n = 36$), the number of possible passwords is $N = 36^1 = 36$. If a password is of two characters ($r = 2$) and we choose from 36 characters ($n = 36$), the number of possible passwords is $N = 36^2 = 1{,}296$. If a password is of six characters ($r = 6$) and we choose from 36 characters ($n = 36$), the number of possible passwords is $N = 36^6 = 2{,}176{,}782{,}336$.

A permutation is a circular permutation if the objects are arranged in the circular form. In such a case the number of possibilities is smaller. The number of circular permutations of n dissimilar things taken r at a time is calculated according to equation:

$$N = \frac{n!}{(n-r)!r} \quad (4.20)$$

Permutations of similar objects are calculated considering the number of objects n and the number of similar objects: $n1$, $n2$, $n3$, $n4$, considering that the sum of all those equals n.

$$N = \frac{(n1 + n2 + n3 + n4)!}{n1!n2!n3!n4!} \tag{4.21}$$

Example The letters from word LETTER can be arranged in many ways in a set of letters. Their number is calculated considering six-letters word, considering that letter L and R appear once and letters E and T appear twice:

$$N = \frac{6!}{1!2!2!1!} = 180 \tag{4.22}$$

4.4.2 Combinations

Combinations without repetition and combinations with repetition are considered.

4.4.2.1 Combinations Without Repetition

If repetition is not allowed, the number of possibilities (N) is calculated considering a number of objects (n) and considering a number of objects in a set (r) according to equation:

$$N = \frac{n!}{(n-r)!r!} \tag{4.23}$$

Example The number of five-card hands possible from a standard 52-card deck is calculated as follows:

$$N = \frac{52!}{(52-5)!5!} = 2,598,960 \tag{4.24}$$

4.4.2.2 Combinations With Repetition

If repetition is allowed, the number of possibilities (N) is calculated considering a number of objects (n) and considering a number of objects in a set (r) according to equation:

$$N = \frac{(n+r-1)!}{(n-1)!r!} \tag{4.25}$$

Example There are five pieces of different fruit to choose from: apple, orange, pear, banana, and mango ($n = 5$), and you choose three of them ($r = 3$). Order does not matter, and you can repeat.

$$N = \frac{(5 + 3 - 1)!}{(5 - 1)!3!} = 35 \tag{4.26}$$

4.5 Conditional Probability and Bayesian Theorem

4.5.1 Conditional Probability

A conditional probability of an event is the probability given that another event has occurred. For example, what is the probability that the total of two dice will be greater than 8 given that the first die is a 6? This can be computed by considering only outcomes for which the first die is a 6. Then, determine the proportion of these outcomes that total more than 8. Figure 4.4 shows all possible outcomes for two dice.

There are 6 outcomes for which the first die is a 6, and of these, there are four that total more than 8 (6,3; 6,4; 6,5; 6,6). The probability of a total greater than 8 given that the first die is 6 is therefore 4/6 = 2/3.

More formally, this probability can be written as: $p(\text{total} > 10|\text{Die1} = 6) = 2/3$.

The expression to the left of the vertical bar represents the event and the expression to the right of the vertical bar represents the condition: the probability that the total is greater than 8 given that die 1 is 6 is 2/3. In general, $p(A|B)$ is the probability of event A given that event B occurred.

$$P(A|B) = \frac{P(A \cap B)}{P(B)} \tag{4.27}$$

If $P(B) = 0$, then the conditional probability $P(A|B)$ is undefined.

The graphical representation of the conditional probability is useful with a probability tree. Figure 4.5 shows an example of probability tree, where four red and one blue coins are in jar. The events of interest are two events of picking coins and not returning them back to the jar.

The probability of picking two red coins equals 0.6. The probability of picking one red and one blue coin equals 0.4. The probability of picking two blue coins equals 0. The conditional probability of picking the red coin in the second selection, if in the first selection was red picked, is 0.75.

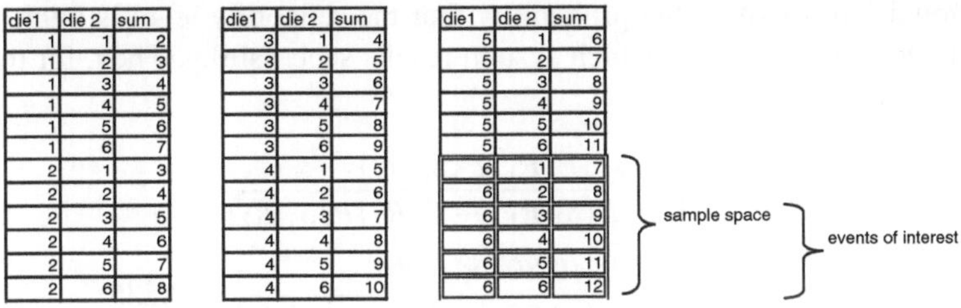

Fig. 4.4 Example of conditional probability (total of two dice, if the first is six)

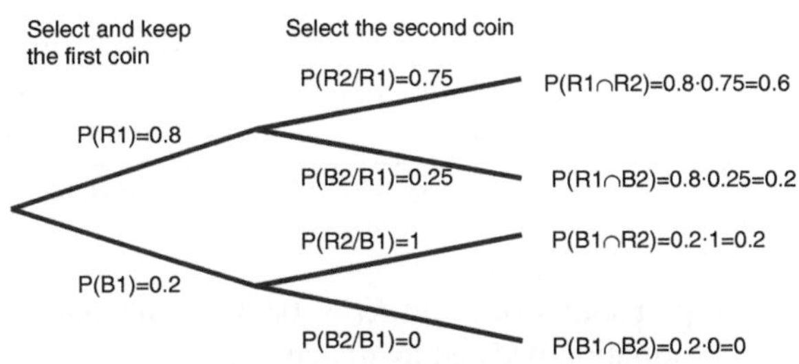

Fig. 4.5 Example of a probability tree

4.5.2 Bayes Theorem

Bayes theorem is a theorem of probability theory, which can be seen as a way of understanding how the probability that a theory is true is affected by a new piece of evidence.

Bayes theorem is stated as posterior probabilities proportional to prior probabilities times likelihoods. A general form is given as follows [10]:

$$P\{A_i|B\} = \frac{P\{A_i, B\}}{P\{B\}} = \frac{P\{A_i\}P\{B|A_i\}}{\sum_i P\{A_i\}P\{B|A_i\}} \tag{4.28}$$

where $A_1, A_2, \ldots, A_n$ is a set of mutually exclusive and exhaustive events ($i = 1, \ldots, n$), $P\{A_i\}$ is the prior probability of A_i before observation, B is the observation, $P\{B|A_i\}$ is the likelihood that the probability of the observation given is true, and $P\{A_i|B\}$ is the posterior probability, i.e., the probability of A_i now that B is known.

The transformation from $P\{A_i\}$ to $P\{A_i|B\}$ is so called Bayes transform. It uses the advantage that the likelihood of $P\{B|A_i\}$ is more easily calculated than $P\{A_i|B\}$.

Example of discrete Bayes theorem discusses the reliability of a new untested system [10]. On the contrary to a classical approach, the Bayes approach treats reliability as a random variable and not as an unknown constant. Based on past experience, we believe that there is an 80% chance that the systems reliability is $R_1 = 0.95$ and 20% chance that the systems reliability is $R_2 = 0.75$. The first system was tested and it resulted with success.

We would like to know the probability that the reliability level is 0.95.

S_i is defined as an event in which system results successfully. Then, for the first success S_1, we get:

$$P\{R_1|S_1\} = \frac{P\{R_1\}P\{S_1|R_1\}}{P\{R_1\}P\{S_1|R_1\} + P\{R_2\}P\{S_1|R_2\}}$$
$$= \frac{(0.80) \times (0.95)}{(0.80) \times (0.95) + (0.20) \times (0.75)} = 0.835 \qquad (4.29)$$

If we assume that a second system was also tested successfully, then we get:

$$\Pr\{R_1|S_1, S_2\} = \frac{\Pr\{R_1\}\Pr\{S_1, S_2|R_1\}}{\Pr\{R_1\}\Pr\{S_1, S_2|R_1\} + \Pr\{R_2\}\Pr\{S_1, S_2|R_2\}}$$
$$\Pr\{R_1|S_1, S_2\} = \frac{(0.80) \times (0.95 \times 0.95)}{(0.80) \times (0.95 \times 0.95) + (0.20) \times (0.75 \times 0.75)} = 0.865 \qquad (4.30)$$

The probability of event $R_1 = 0.95$ was updated by applying Bayes theorem as new information became available.

4.6 Random Variables

A random variable is any variable determined by chance and with no predictable relationship to any other variable. The selection of the random variable is unpredictable and cannot be subsequently reproduced. A random variable generator is a tool for generation of random variables. Actually, two methods are known to get the random variables:

- Measuring a physical phenomenon that is expected to be random.
- Running a pseudorandom variable generator that is a computational algorithm determined by a seed.

Random variable is a more general term and random number is a subset term, which is commonly used if the variable under investigation is a number.

Figure 4.6 shows two examples of generating a set of real random numbers between 0 and 1.50, and 1,000 generated real random numbers between 0 and 1 are presented on both portions of the figure, respectively. The comparison of two portions shows that the generated real numbers in larger quantities enables higher degree of our confidence that the generation is really random. But only the subjective view to the figures is not enough.

Table 4.1 shows the resulted outcomes of 600 tosses of one square six-sided die. Each number in a table represents the outcome of one toss of one die. If the die is a fair die, one should expect that numbers 1, 2, 3, 4, 5, and 6 appear randomly,

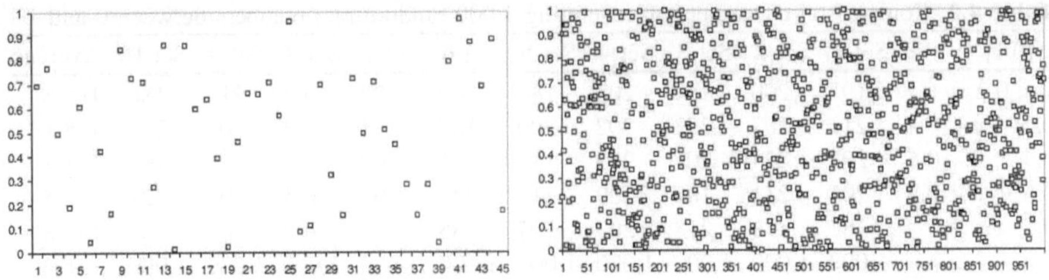

Fig. 4.6 Examples of a real random number between 0 and 1 (50 and 1,000 numbers)

Table 4.1 Example of results of 600 tosses of one square six-sided die

2, 6, 2, 1, 5, 6, 5, 6, 6, 6, 5, 4, 4, 6, 1, 5, 3, 4, 6, 5, 1, 5, 6, 5, 2, 2, 4, 3, 2, 3, 2, 1, 4, 3, 2, 3, 6, 6, 6, 2, 5, 3, 5,
6, 4, 5, 2, 3, 3, 1, 4, 6, 3, 4, 2, 2, 5, 5, 2, 1, 1, 6, 6, 1, 1, 5, 3, 2, 2, 3, 6, 3, 4, 2, 2, 6, 6, 3, 5, 1, 6, 3, 3, 4, 5, 3,
6, 5, 6, 1, 5, 4, 2, 5, 6, 1, 2, 3, 3, 1, 6, 4, 6, 4, 1, 6, 1, 1, 6, 3, 3, 4, 3, 1, 1, 3, 6, 4, 3, 5, 3, 3, 6, 3, 1, 3, 1, 6, 1,
6, 5, 3, 5, 3, 1, 4, 6, 6, 2, 4, 4, 6, 1, 3, 1, 1, 3, 1, 2, 4, 3, 5, 3, 1, 1, 5, 6, 3, 5, 2, 5, 1, 3, 4, 2, 1, 4, 1, 1, 2, 3, 4,
4, 4, 2, 1, 2, 5, 5, 2, 2, 4, 6, 5, 1, 2, 5, 4, 1, 4, 4, 5, 4, 2, 3, 6, 1, 1, 2, 3, 3, 5, 1, 2, 2, 2, 6, 5, 2, 4, 5, 5, 6, 5, 1,
2, 5, 4, 2, 2, 2, 5, 6, 4, 5, 6, 6, 2, 4, 6, 2, 3, 5, 6, 6, 5, 2, 5, 2, 2, 3, 4, 2, 1, 3, 6, 6, 2, 3, 5, 4, 3, 2, 6, 4, 5, 4, 4,
4, 1, 2, 2, 1, 3, 3, 2, 2, 3, 5, 2, 1, 6, 5, 2, 1, 4, 1, 4, 1, 5, 5, 2, 6, 4, 1, 5, 3, 6, 5, 3, 4, 5, 4, 1, 4, 1, 1, 6, 6, 6, 5,
6, 4, 3, 6, 5, 1, 4, 6, 4, 5, 5, 3, 2, 2, 4, 2, 4, 6, 4, 3, 4, 5, 2, 6, 2, 4, 1, 6, 4, 1, 1, 1, 5, 4, 5, 5, 6, 6, 1, 6, 1, 4, 6,
2, 5, 4, 4, 6, 3, 1, 6, 3, 3, 4, 4, 1, 5, 5, 1, 2, 6, 6, 3, 3, 6, 4, 1, 4, 1, 5, 2, 3, 5, 1, 1, 2, 3, 5, 2, 5, 6, 2, 1, 4, 6, 5,
4, 5, 1, 6, 3, 5, 2, 5, 3, 1, 4, 5, 3, 5, 5, 1, 2, 3, 6, 5, 6, 1, 3, 1, 1, 6, 6, 6, 6, 1, 2, 3, 1, 1, 3, 3, 2, 6, 5, 1, 1, 6, 6,
3, 6, 6, 2, 4, 2, 5, 1, 6, 2, 4, 6, 5, 5, 6, 5, 5, 1, 5, 6, 1, 1, 5, 1, 6, 4, 2, 6, 2, 6, 3, 2, 5, 3, 3, 2, 4, 3, 6, 1, 2, 4, 2,
6, 4, 4, 1, 6, 3, 4, 4, 5, 6, 3, 2, 2, 3, 3, 3, 6, 5, 4, 3, 6, 1, 6, 3, 3, 4, 5, 3, 6, 1, 4, 1, 4, 6, 1, 5, 3, 4, 6, 5, 2, 1, 5,
1, 2, 2, 3, 6, 3, 1, 6, 2, 4, 5, 5, 5, 5, 3, 1, 1, 1, 6, 4, 5, 4, 1, 6, 1, 6, 6, 4, 5, 3, 2, 6, 5, 4, 1, 6, 3, 1, 2, 1, 4, 2, 6,
4, 1, 1, 6, 3, 3, 4, 5, 4, 5, 4, 4, 3, 5, 4, 1, 1, 5, 3, 6, 3, 1, 3, 2, 3, 6, 5, 5, 1, 1, 5, 3, 1, 6, 1, 3, 3, 2, 1, 3, 2

and after the large number of tosses, the number of the outcomes is equal and approximately equal to 1/6 of the number of tosses.

The most difficult is to prove that the measures of the physical phenomenon give really a set of random numbers or that running a number generator gives really a set of random numbers. The simplest tests can be done in a way to group the outcomes of random experiments into the groups and evaluate the portion of the number of the occurrences within one group versus the all number of occurrences. A large number of generated random numbers is a prerequisite for proving the randomness of the pseudorandom number generator.

The grouping of the real numbers generated at the example of generation of 1,000 random real numbers between 0 and 1 can be, for example, a grouping of a generated random numbers in the intervals: $x \leq 0.2$; $0.2 \leq x < 0.4$; $0.4 \leq x < 0.6$; $0.6 \leq x < 0.8$; $0.8 \leq x$. Table 4.2 shows the number of generated random numbers grouped in five such groups for ten sets of generated random numbers. It is expected that 200 of generated numbers are in one such group, which is the theoretical average. Table 4.2 shows that the average of the defined groups well matches with the expected value, which increases our confidence that the generation of random numbers is really random.

Table 4.2 Comparison of ten sets of generating 1,000 random real numbers between 0 and 1

x [0.1]	Set 1	Set 2	Set 3	Set 4	Set 5	Set 6	Set 7	Set 8	Set 9	Set 10	Average
$x < 0.2$	197	200	184	203	183	202	222	204	217	186	199.8
$0.2 < x < 0.4$	211	187	203	202	209	187	181	194	200	207	198.1
$0.4 < x < 0.6$	203	224	207	203	188	201	205	211	181	195	201.8
$0.6 < x < 0.8$	201	211	188	189	213	225	197	185	193	223	202.5
$0.8 < x$	188	178	218	203	207	185	195	206	209	189	197.8
	1,000	1,000	1,000	1,000	1,000	1,000	1,000	1,000	1,000	1,000	1,000

Table 4.3 Comparison of four sets of results of generating the 600 tosses of one square six-sided die

	Set 1	Set 2	Set 3	Set 4	Average
Toss outcome 1	104	91	104	96	98.75
Toss outcome 2	100	93	84	107	96
Toss outcome 3	104	103	92	95	98.5
Toss outcome 4	91	107	110	86	98.5
Toss outcome 5	115	101	107	116	109.75
Toss outcome 6	86	105	103	100	98.5
Sum	600	600	600	600	600

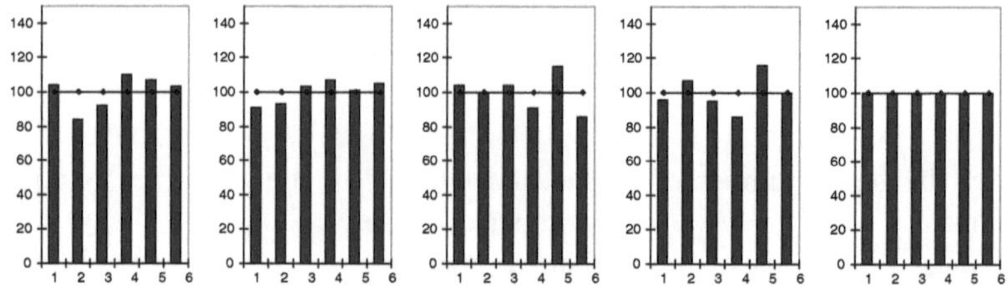

Fig. 4.7 Comparison of four sets of results of generating the 600 tosses of one square six-sided die

The grouping of the generated tosses of a die can be done regarding the toss outcome. Table 4.3 and Fig. 4.7 show four sets of results of generating the 600 tosses, where it is observed that the die is relatively fair and that the number of outcomes is approximately as expected 100 per each of the outcomes, which is the average. Figure 4.7 shows four sets of results and the ideal set of results that would mean all six outcomes exactly 100 times. The average of those groups well matches with the expected value, which increases our confidence that the tossing a die is really random.

But such simple tests are not enough for more serious calculations. Figure 4.8 shows an example of a real random number between 0 and 10 (1,000 numbers). From the figure itself, one can notice that only selected values appear many times and many of the values do not appear at all. Furthermore, such an example would

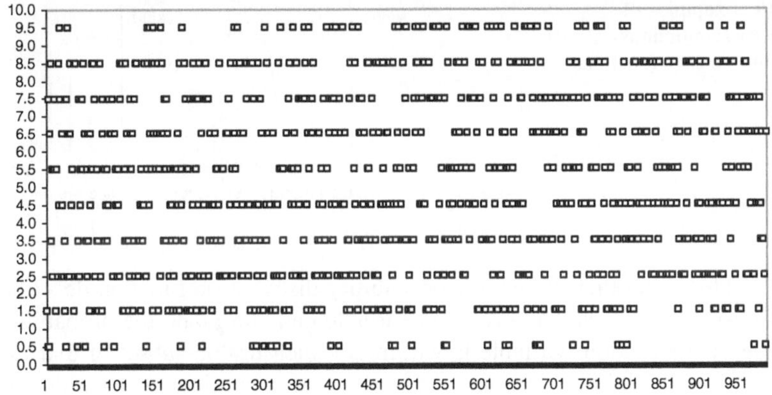

Fig. 4.8 Example of a real random number between 0 and 10 (1,000 numbers)

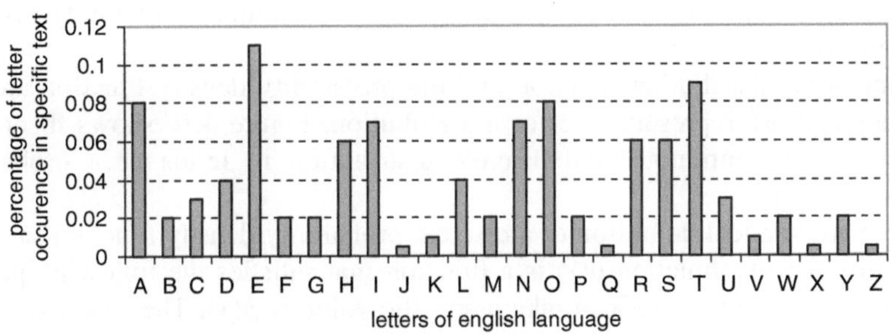

Fig. 4.9 Distribution of percentages of letter occurrence in specific text in English language

pass well a simple test of a grouping of random numbers in the intervals: $x \leq 2$; $2 \leq x < 4$; $4 \leq x < 6$; $6 \leq x < 8$; $8 \leq x$, but would not pass well another test, if the quarter of that interval would be taken as a basis in another test.

If the example of interest is more complex than tossing a coin or die, the term distribution becomes important. For example, one should guess, what is the probability, that a selected letter from a text is a letter a, for example [12, 13]. Figure 4.9 shows that the probability, that a letter a is selected from letters of text, is approximately 8%, if this figure is treated as a representative source of information.

4.7 Probability Distributions

Distribution is a degree to which the outcomes of events are evenly spread over the possible values [2–10, 14–25]. Probability distribution function is a function that represents probabilities to which the outcomes of events are spread over the

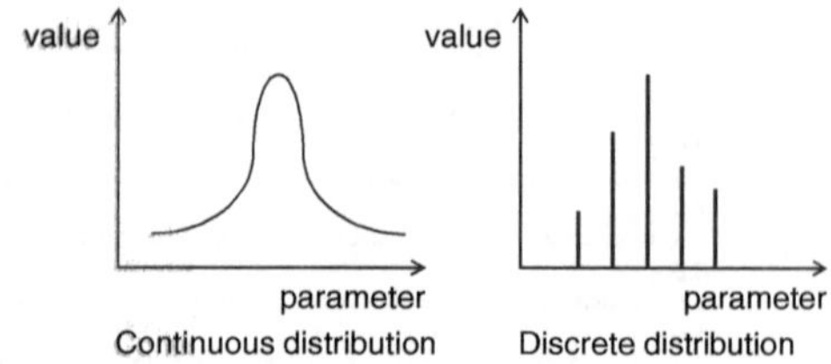

Fig. 4.10 Examples of discrete and continuous probability density functions

possible values. In other words, the probability distribution function describes the range of possible values that a random variable can attain and the probability that the value of the random variable is within any measurable subset of that range.

Probability distribution function can be discrete or continuous depending on the nature of the events that are considered. If a random variable under investigation is a discrete variable, its probability distribution function is a discrete probability distribution function. If a random variable under investigation is a continuous variable, its probability distribution function is a continuous probability distribution function.

Cumulative distribution function and the probability density function are two functions that are representative for a distribution. Figure 4.10 shows theoretical comparison of continuous and discrete distribution in terms of a probability density function.

The mathematical definition of a discrete probability density function says that probability density function $p(x)$ is a function that satisfies the following properties. The probability that x can take a specific value is $p(x)$. The function $p(x)$ is non-negative for all real x.

$$p(x) \geq 0 \tag{4.31}$$

The sum of all probabilities of the probability density function equals 1.

$$\sum_{i=1}^{n} p(x_i) = 1 \tag{4.32}$$

The mathematical definition of a continuous probability density function says that the probability density function $f(x)$ is a function that satisfies the following properties. The probability that x can take a specific value is $f(x)$. The function $f(x)$ is non-negative for all real x.

$$f(x) \geq 0$$

The integral of the probability density function equals one.

$$\int_{-\infty}^{\infty} f(x)dx = 1 \tag{4.33}$$

The probability that x lays between two points a and b is calculated by integrating the probability density function over the interval between a and b.

$$p(a \le x \le b) = \int_{a}^{b} f(x)dx \tag{4.34}$$

The continuous probability density functions are defined for an infinite number of points over a continuous interval and the probability at a single point is zero. The probabilities are measured over intervals and the area under the curve between two distinct points defines the probability for that interval. The height of the probability density function can in theory for a small interval be greater than one. The property of the continuous distribution that the integral must equal one is equivalent to the property for discrete distributions that the sum of all the probabilities must equal one.

In general, the discrete probability functions are referred to as probability mass functions, and continuous probability functions are referred to as probability density functions. The term probability function covers both discrete and continuous distributions. Sometimes, when we refer to a probability function in generic terms, we may use the term probability density functions to mean both discrete and continuous probability functions.

When the random variable takes values in the set of real numbers, the distribution is completely described by the cumulative distribution function, whose value at each real x is the probability that the random variable is smaller than or equal to x.

The relation between cumulative distribution function $F(x)$ and the probability density function $f(x)$ is the following for discrete variables.

$$F(x) = \sum_{x_i \le x} p(x_i) \tag{4.35}$$

The relation between cumulative distribution function $F(x)$ and the probability density function $f(x)$ is the following for continuous variables.

$$f(x) = \frac{dF(x)}{dx} \tag{4.36}$$

$$F(x) = \int_{-\infty}^{x} f(x)dx \tag{4.37}$$

$$\lim_{x \to -\infty} F(x) = 0 \tag{4.38}$$

$$\lim_{x \to \infty} F(x) = 1 \tag{4.39}$$

The probability that x lays between two points a and b can be calculated by the difference of cumulative distribution function at point b and a, or by integrating the probability density function over the interval between a and b.

$$p(a \leq x \leq b) = \int_a^b f(x)dx = F(b) - F(a) \tag{4.40}$$

Probability distributions are used in theory and in practice. Theoretical part is very mathematical. Practical applications are numerous. There are various probability distributions used in various different applications. One of the more important ones is the normal distribution, which is also known as the Gauss distribution or the bell curve and approximates many different naturally occurring distributions. The list of probability distributions includes the following distributions:

- Normal distribution or Gauss distribution or Bell curve
- Lognormal distribution
- Beta distribution
- Gamma distribution
- Uniform distribution
- Weibull distribution
- Exponential distribution
- Binomial distribution
- Poisson distribution
- Delta distribution

Other distributions include the following distributions: χ^2 distribution, inverted gamma distribution, inverted χ^2 distribution, Student distribution, F distribution, and Dirichlet distribution.

Before defining the features of each distribution, let us define some background terms that are important in statistics.

The arithmetic mean of a set of values is the quantity commonly called the average or the mean value. From a set of n discrete samples $x_1, x_2, \ldots, x_n$, the arithmetic mean is calculated as:

$$\text{Mean} = \frac{1}{n}\sum_{i=1}^{n} x_i \tag{4.41}$$

From a continuous function $f(x)$ over the interval [a, b], the arithmetic mean is calculated as:

$$\text{Mean} = \frac{1}{b-a}\int_a^b f(x)dx \tag{4.42}$$

The geometric mean is the quantity calculated from a set of n discrete samples $\{x_1, x_2, \ldots, x_n\}$ as:

$$\text{Geometric mean} = \sqrt[n]{\prod_{i=1}^{n} x_i} \tag{4.43}$$

Mean deviation is one of the most widely used measures of dispersion. More specifically, it is used to indicate the degree to which a given set of data tend to spread about the mean value. If the data are more spread apart, the deviation is the higher.

$$\text{Mean deviation} = \frac{1}{n}\sum_{i=1}^{n} |x_i - \text{mean}| \tag{4.44}$$

A confidence interval is an interval in which a measurement or trial falls corresponding to a given probability. It is determined by a particular confidence level, usually expressed as a percentage. The confidence limits are the end points of the confidence interval.

The median is the value of the point that has half the data smaller than that point and half the data larger than that point. If $x_1, x_2, \ldots, x_n$ is a random sample sorted from the smallest value to a largest value, then the median is defined as:

$$\text{Median} = M = \begin{cases} x_{(n+1)/2} & \text{for odd } n \\ \frac{\left(x_{n/2} + x_{(n/2)+1}\right)}{2} & \text{for even } n \end{cases} \tag{4.45}$$

Or, for continuous distribution, the integral of intervals before and after the median value are equal to ½.

$$\int_{-\propto}^{\text{median}} f(x)dx = \frac{1}{2} = \int_{\text{median}}^{\propto} f(x)dx \tag{4.46}$$

The median is a particular case of α percentile, where $\alpha = 50\%$.

Mid-range is simply the average of maximal and minimal value.

The expected value of a random variable X is defined by equation

$$E[X] = \sum_{x:p(x)>0} xp(x) \tag{4.47}$$

Let X be a discrete random variable with probability mass function $p(x)$. For any function $g(x)$, the expected value is defined by equation, where the sum is obtained for the definition space $D(p)$.

$$E[g(X)] = \sum_{x \in D(p)} g(x)p(x) \tag{4.48}$$

Variance is a measure of the dispersion of a set of data points around their mean value. Variance is a mathematical expectation of the average squared deviations from the mean. The variance is the square of the standard deviation.

$$\text{variance} = \frac{1}{n}\sum_{i=1}^{n} (x_i - \text{mean})^2 \tag{4.49}$$

The variance of the random variable X with a finite mean μ is defined by equation

$$\text{Var}(X) = E\left[(X - \mu)^2\right] \tag{4.50}$$

$$\text{Var}(X) = E[X^2] - (E[X])^2 \tag{4.51}$$

Standard deviation is a measure of the dispersion of a set of data from its mean. If the data are more spread apart, the deviation is higher. Standard deviation is calculated as the square root of variance.

$$\text{Standard deviation} = \sigma = \sqrt{\frac{1}{n}\sum_{i=1}^{n} (x_i - \text{mean})^2} \tag{4.52}$$

The standard deviation of a random variable X is defined by equation

$$\text{SD}(X) = \sqrt{\text{Var}(X)} \tag{4.53}$$

4.7.1 Normal Distribution or Gauss Distribution or Bell Curve

The Gauss distribution is the most common probability distribution in science. Repeated independent measurements with random uncertainties of almost any quantity follow this distribution. For example, the numbers of heads and tails you are likely to find if you flip a coin many times are described by a Gauss distribution. It is a symmetric function round its mean value. Gauss distribution is derived from binomial distribution considering that the number of repetitions is very high. The probability density function is defined for x from $-\infty < x < \infty$ given by equation.

$$f(x) = \frac{1}{\sigma\sqrt{2\pi}} \times e^{-\frac{(x-\mu)^2}{2\sigma^2}} \tag{4.54}$$

where σ is the standard deviation of variable x, and μ is the mean value of variable x.

$$E[X] = \mu \tag{4.55}$$

$$\text{var}(X) = \sigma^2 \tag{4.56}$$

4.7.2 Lognormal Distribution

Lognormal distribution indicates that the logarithm of the random variable x is normally distributed. The probability density is defined for x from $0 < x < \infty$ given by equation.

$$f(x) = \frac{1}{\sigma x \sqrt{2\pi}} \times e^{-\frac{(\ln(x)-\mu)^2}{2\sigma^2}} \tag{4.57}$$

The scale parameter μ of the normal and lognormal distributions equals to the mean (and median) value for the normal distribution.

$$\sigma = \frac{\ln(\text{EF})}{1.6449} \tag{4.58}$$

The shape parameter σ of the normal and lognormal distributions equals to the standard deviation for the normal distribution.

$$M = e^{\mu} \tag{4.59}$$

$$E(X) = e^{\left(\mu + \frac{\sigma^2}{2}\right)} \tag{4.60}$$

4.7.3 Beta Distribution

$$f(x) = \frac{\Gamma(\alpha+\beta)}{\Gamma(\alpha)\Gamma(\beta)} x^{\alpha-1}(1-x)^{\beta-1} \tag{4.61}$$

$$E(X) = \frac{\alpha}{\alpha+\beta} \tag{4.62}$$

where α is the scale parameter and β is the shape parameter.

4.7.4 Gamma Distribution

Gamma distribution is defined as follows.

$$f(x) = \frac{\beta^{\alpha}}{\Gamma(\alpha)} x^{\alpha-1} e^{-\beta x} \tag{4.63}$$

$$E(X) = \frac{\alpha}{\beta} \tag{4.64}$$

where α is the scale parameter and β is the shape parameter.

4.7.5 Uniform Distribution

Uniform distribution is defined for values between two numbers; for others, it equals zero.

$$f(x) = \frac{1}{x_{\max} - x_{\min}} \tag{4.65}$$

$$E(X) = \frac{x_{\max} + x_{\min}}{2} \tag{4.66}$$

4.7.6 Binomial Distribution

A random variable k $\{0, 1, 2, \ldots, n\}$ with the following probability mass function is called a binomial random variable with parameters n and p

$$p(k) = P(X = k) = \binom{n}{k} p^k (1 - p)^{n-k} \tag{4.67}$$

$$E(X) = n \times p \tag{4.68}$$

$$\mathrm{Var}(X) = n \times p \times (1 - p) \tag{4.69}$$

4.7.7 Poisson Distribution

Poisson distribution is defined for a random variable k $\{0, 1, 2, \ldots, n\}$ and $\lambda \geq 0$ is defined by equation

$$p(k) = e^{-\lambda} \frac{\lambda^k}{k!} \tag{4.70}$$

$$E[X] = \mathrm{var}(X) = \lambda \tag{4.71}$$

4.7.8 Delta Function Distribution

The simplest distribution that will be considered is the delta function distribution. The delta function is a mathematical tool that causes a continuous variable to have only one possible value.

$$\delta(x)\,dx = \begin{cases} 1 & \text{for } x = 0 \\ 0 & \text{otherwise} \end{cases} \tag{4.72}$$

For a delta function, there is no variability of the parameter x. Delta functions reduce integrals to single values.

4.7.9 Weibull Distribution

The Weibull distribution is defined by equation

$$f(t) = \frac{\beta}{\mu} \times \left(\frac{t-\gamma}{\mu}\right)^{\beta-1} \times e^{\left[-\left(\frac{t-\gamma}{\mu}\right)^{\beta}\right]} \tag{4.73}$$

where β is the shape parameter μ is the scale parameter and γ is the location parameter.

$t \geq \gamma$, $\beta \geq 0$, $\mu \geq 0$

The Weibull distribution for $\beta = 1$ becomes exponential distribution. The Weibull distribution for $\beta < 1$ becomes gamma distribution. The Weibull distribution for $\beta = 2$ becomes lognormal distribution.

4.7.10 Exponential Distribution

The exponential distribution is widely used for modeling time to failure where $t \geq 0$.

$$f(t) = \lambda \times e^{-\lambda t} \tag{4.74}$$

$$E[X] = \frac{1}{\lambda} \tag{4.75}$$

$$\mathrm{Var}(X) = \frac{1}{\lambda^2} \tag{4.76}$$

The exponential distribution is an excellent model for the long flat portion of the bathtub curve, because of its constant failure rate property. Most components and systems spend most of their lifetimes in flat portion of the bathtub curve, which justifies frequent use of the exponential distribution. The use of exponential distribution is useful when early failures or wear out failures are not specially considered.

4.8 Bathtub Curve

The bathtub failure rate concept is widely used to represent failure behavior of many engineering items. The term bathtub stems from the fact that the shape of the failure rate curve resembles a bathtub. The bathtub curve consists of three periods:

- An infant mortality period with a decreasing failure rate
- A normal life period or useful life period with a low and relatively constant failure rate
- A wear-out period that exhibits an increasing failure rate

Figure 4.11 shows schematically the bathtub curve. The infant mortality period is mostly caused by defects designed into or built into a product. The product manufacturer determines the methods to eliminate the defects with appropriate specifications, robustness, and adequate design tolerance.

The normal life period is far longer period than the other two. This is not observed from the figure. The product is in operation and subjected to regular testing and maintenance, if applicable.

The wear-out period starts when the degradation of the product because of ageing and wear-out increases. The wear-out period can be prolonged by additional maintenance until this is not too costly.

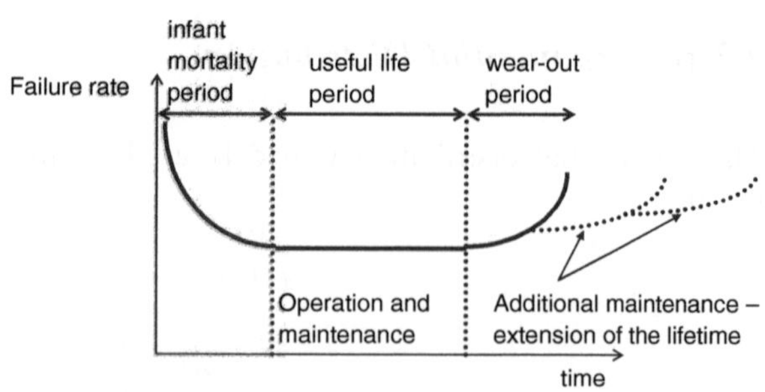

Fig. 4.11 Bathtub curve

References

1. Kolmogorov A (1993) Grundbegriffe der Warscheinlichkeitsrechnung. Springer Verlag, Berlin
2. Shiryaev AN (1996) Probability. Springer, New York
3. Bertsekas DP, Tsitsiklis JN (2002) Introduction to probability. Athena Scientific, Belmont, MA
4. Rychlik I, Rydén J (2006) Probability and risk analysis: an introduction for engineers. Springer, New York
5. Durrett R (1991) Probability: theory and examples. Wadsworth-Brooks/Cole, Pacific Grove, CA
6. Tijms H (2007) Understanding probability: chance rules in everyday life. Cambridge University Press, Cambridge
7. Feller W (1968) An Introduction to probability and its applications. Wiley, New York
8. Chung KL (1974) Elementary probability with stochastic processes. Springer, New York
9. Vesely W, Dugan J, Fragola J et al (2002) Fault tree handbook with aerospace applications. National Aeronautics and Space Administration
10. Kumamoto H, Henley EJ (1996) Probabilistic risk assessment and management for engineers and scientists. IEEE, New York
11. Weisstein E, Pascal's triangle: From MathWorld-A Wolfram Web Resource. http://mathworld.wolfram.com/PascalsTriangle.html. Accessed 4 Aug 2010
12. Letter frequencies. http://www.simonsingh.net/The_Black_Chamber/frequencyanalysis.html. Accessed 24 Aug 2010
13. NIST/SEMATECH e-handbook of statistical methods. http://www.itl.nist.gov/div898/hand book/. Accessed 24 Aug 2010
14. Atwood CL, La Chance JL, Martz HF et al (2003) Handbook of parameter estimation for probabilistic risk assessment (NUREG/CR-6823). Nuclear Regulatory Commission
15. MIL-HDBK-338 (1984) Electronic reliability design handbook
16. Dhilon BS (2007) Applied reliability and quality. Springer, London
17. Bain LJ, Engelhardt M (1992) Introduction to probability and mathematical statistics. PWS-Kent, Boston
18. Box GEP, Tiao GC (1973) Bayesian inference in statistical analysis. Addison-Wesley, Reading
19. Çinlar E (1975) Introduction to stochastic processes. Prentice-Hall, Englewood Cliffs, NJ
20. Derman C, Gleser LJ, Olkin I (1973) A guide to probability theory and application. Holt, Rinehart and Winston, New York
21. Meester R (2008) A natural introduction to probability theory. Birkhäuser Basel, Boston, MA
22. Geiss C, Geiss S (2009) An introduction to probability theory. http://users.jyu.fi/~geiss/scripts/introduction-probability.pdf. Accessed 27 July 2010
23. Williams D (1991) Probability with martingales. Cambridge University Press, Cambridge
24. Billingsley P (1995) Probability and measure. Wiley, New York
25. Bauer H (2001) Measure and Integration Zheory. Walter de Gruyter, Berlin

Part II
Reliabilty Methods

Chapter 5
Fault Tree Analysis

The first fault tree in the history was the one from which Eve took the forbidden apple in the Garden of Eden

Lee Remick

Probabilistic Safety Assessment conference, 1993

5.1 Introduction

The fault tree analysis is a standard method for the assessment and improvement of reliability and safety [1–3]. It has been and it is applied in various sectors, such as nuclear industry, air and space industry, electrical industry, chemical industry, railway industry, transport, software reliability, and insurance. Its widely acceptance is gained primarily when integrated with the event tree analysis as a part of the probabilistic safety assessment (PSA) for improving the safety of nuclear power plants and for improving the safety of space missions [4–7].

The fault tree analysis is an analytical technique, where an undesired state of the system is specified and then the system is analyzed in the context of its environment and operation to find all realistic ways in which the undesired event can occur. The undesired state of the system, which is identified at the beginning of the fault tree analysis, is usually a state that is critical from a safety or reliability standpoint and is identified as the top event. Top event is therefore an undesired event, which is further analyzed with the fault tree analysis.

The fault tree analysis is a term that combines the graphical model, which is called the fault tree or fault tree model, the qualitative analysis of the fault tree, and the quantitative analysis of the fault tree, which includes the probabilistic failure data and the associated results.

The fault tree is a graphic model of the various parallel and sequential combinations of faults that can lead to the occurrence of the predefined undesired event or top event.

The logical gates of the fault tree integrate the primary events to the top event. The primary events are the events that are not further developed, e.g., the basic events and the house events.

The basic events are the ultimate parts of the fault tree, which represent the undesired events and their failure modes, e.g., the component failures, the missed actuation signals, the human errors, the unavailabilities because of the test and

M. Čepin, *Assessment of Power System Reliability*,
DOI: 10.1007/978-0-85729-688-7_5, © Springer-Verlag London Limited 2011

maintenance activities, the common cause failure contributions, and software errors.

The house events represent the conditions set to either true or false, which support the modeling of connections between the gates and the basic events and enable that the fault tree better represents the system operation and its environment.

5.2 Fault Versus Failure

Failures are specific events which are the outcomes of the failure modes, and are directly connected with reasons for failure within the boundaries of the equipment under investigation. Faults are more general events and consider the behavior outside of defined equipment boundaries in addition. If a relay does not close, because the contacts are broken, this is a failure of the relay. If the relay does not close, because the signal for relay closure is not received, it is not the failure of the relay. But as the relay cannot function and it should, such state is its fault.

Another example of distinction of definitions of failure and fault is that the switch closes at the wrong time because of the improper functioning of some component. This is not a switch failure, but the circuit can enter into an unsatisfactory state or fault state. To generalize, all failures are faults but not all faults are failures.

The fault tree is not a model of all equipment faults. It is a model of only those faults or of those failure modes, which can cause the top event to occur.

Fault tree is a static tool, although it has evolved to a number of semi-dynamic or dynamic attempts.

5.3 Fault Tree Analysis Procedure Steps

The fault tree analysis procedure steps are the following [5, 8–12].

- Identification of the objectives for the fault tree analysis
- Definition of the top event of the fault tree
- Definition of the scope, resolution, and ground rules of the fault tree
- Fault tree construction
- Qualitative fault tree evaluation
- Preparation of the probabilistic failure database and connection of the basic events of the fault tree with probabilistic failure data
- Quantitative fault tree evaluation
- Interpretation of the fault tree analysis results

Figure 5.1 shows the fault tree analysis procedure steps together with their main mutual relationships. The procedure steps are described in the following subsections.

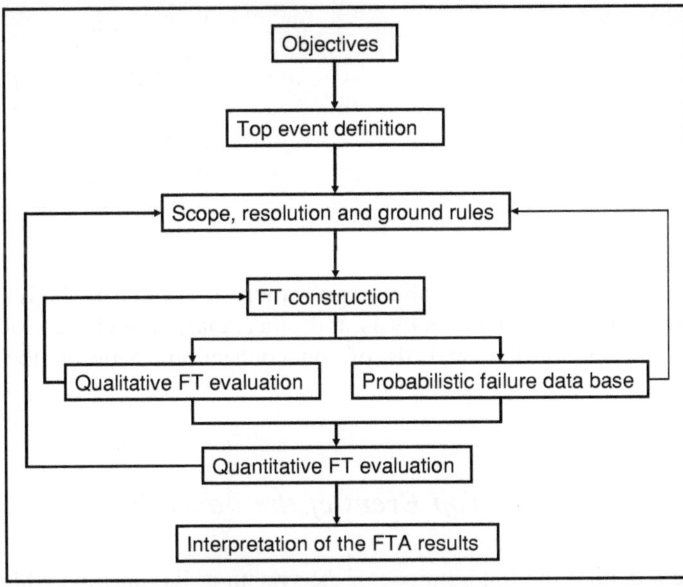

Fig. 5.1 Fault tree procedure steps

5.3.1 Objectives For the Fault Tree Analysis

One of more objectives can be defined, which can vary for the case of existing and operating system or facility or for the case of system or facility in the conceptual or design stage. The most common objectives can include one or more objectives from the following list:

- Assessment of the failure probability of the system or the system function (or assessment of the reliability of the system or availability or their complements: either unreliability or unavailability)
- Comparison of the variations of the system design
- Fulfillment of the regulatory objectives
- Identification of the most important components of the system in terms of its reliability
- Identification of the most important components of the system in terms of maintenance priority determination
- Improvement of the documentation of the system and maintaining the knowledge about its behavior

The objectives should be defined in terms of undesired function or functions of the system. Mostly, the systems of interest are connected with the higher level of the analysis, where the connections between systems are considered in addition, which is realized through the probabilistic safety assessment [11–15]. Probabilistic

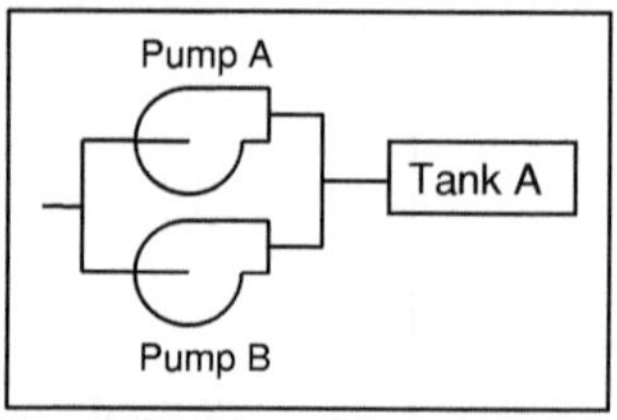

Fig. 5.2 Simple example system

safety assessment is a term describing application of several methods including fault tree analysis for systems analysis and other analysis methods such as the event tree analysis for the analysis of the connections between the systems [16–23].

5.3.2 Definition of the Top Event of the Fault Tree

The top event defines the failure mode of the system or its function, which is then analyzed in terms of failure modes of its components and influence factors. It is important to distinguish the success criteria of the system and the description of the top event, which is defined in term of failure modes [5, 6, 11].

This distinction between success criteria of the system and description of the top event is shown for the simple example system. Figure 5.2 shows the example system.

The simple example system consists of two parallel pumps and one tank. Note that pumps are usually accompanied with valves on the same pipeline but here the valves are intentionally neglected to present a simple example. The function of the system requires such capacity of the water to be delivered to the tank for 4 h that one pump is capable enough. Therefore, the success criterion for the system is operation of one out of two pumps for 4 h. In other words, at least one of the pumps has to be running for 4 h in order that the system works as required. The initial state of system components is such that the tank is empty and the pumps are stopped.

If the objective of the fault tree analysis would be to assess the reliability of the system, the top event would be defined as failure of two out of two pumps to operate for 4 h. In other words, both pumps have to fail running for 4 h in order that the system fails.

Then, failure of the system is defined as the failure to satisfy the given success criteria.

If the objective of the fault tree analysis would be to assess the reliability of the system, which consists of 7 parallel protection lines and proper function of 2 of them is needed for the success of the system, the success criterion would be 2 out of 7 or written differently: 2/7. The top event in the fault tree would be then defined accordingly: if the proper function of 2 parallel protection lines is sufficient for proper operation of the system, then the failure of 6 lines would be needed in order that the system has failed. So the top event in the fault tree, which represents the system failure, would be defined as failure of 6 out of 7 lines or written as 6/7.

In general, the success criteria of certain system can be defined as k out of n subsystems or written: k/n, which means that n parallel subsystems exist and k of them are needed for proper system function. The top event in the fault tree, which represents the system failure, can be defined as $(n - k + 1)/n$. This means that $n - k + 1$ parallel lines have to fail to have the system failure.

One interesting special case is the logic 2 out of 3, which means that two out of three parallel lines have to operate properly for the proper system operation. If we determine, what should be defined in the top event of the fault tree, we can calculate $n - k + 1/n$, which is $(3 - 2 + 1)/3$, which is again two out of three. This special case is clearly logical. If we have three parallel lines of which two lines have to operate properly in order that the system operates, the failure of the system is reached if two out of three parallel lines fail.

An example of a top event when analyzing safety systems in nuclear power plants would be: high pressure safety injection system fails to deliver water (2 out of 2 lines has to fail) to reactor coolant system, because the function of one line is enough for the system to operate properly.

An example of a top event when analyzing safety of space missions would be "loss of a space vehicle." Or another example "loss of crew." An example of the top event when calculating the reliability of a power substation could be loss of power to the load A or loss of power to the specific set of customers.

5.3.3 Definition of the Scope, Resolution, and Rules of the Fault Tree

The scope of the fault tree indicates which of the faults and contributors are included in the model and subsequently in the analysis and which are not.

The state of the analyzed facility is frozen in terms of current design revision at a specified date. If the design changes of the facility are made or if the procedures changes appear during the duration of the analysis project, they are not considered immediately. If needed, the new frozen date and design revision is determined and the revision of the models is then made.

The scope includes that the boundary conditions of the analysis are defined. The boundary conditions include the way how treatment of the outside events is considered in the analysis and where the limits of the system are. For example of a simple example system from Fig. 5.2, it could be assumed that electrical power is available and its loss is not considered or alternatively the loss of power supply to each of the pumps is further analyzed. Or the actuation signal for the pumps start could be considered or not as a means of success or failure of the system. In addition, the initial states of the components are defined. If the pumps are stopped, they can fail to start and then can latter fail to run, but if the pumps would be running, the only failure mode considered in the fault tree would be failure to run for required time.

The resolution of the fault tree analysis determines how large the components are. For example, the diesel generator can be considered as a component and it is not further divided in the analysis. The basic events of the fault tree are then related to the failure modes of the diesel generator as a whole and they are only a few. Or alternatively, the pieces of the diesel generator can be assumed as the components such as the daily tank, the instrumentation, the rotor, the stator and the associated cables and pipes for example. Then, the basic events of the fault tree are related to the larger number of components of the diesel generator and their failure modes.

The resolution of modeling is closely related to the reliability database. The resolution is selected such that the components of the system suit the items from the reliability database that it is used.

The ground rules for the fault tree analysis include the procedures how the fault tree is developed and what are the rules for determining the descriptions and their abbreviations.

The fault tree can be developed in sense that the system and its components are followed sequentially one after another through the paths of the system in a way that the gates have only two inputs: (i) particular component failure mode or (ii) the input to that component. Or the fault tree gates are enabled with more input events or even the portions of the system, which include several components that are treated as modules that are then jointly considered within the fault tree.

An important feature of the analysis of more complex facilities with several systems, several functions, and a large number of components is the naming scheme of the basic events and gates. This can at a later stage largely simplify the difficulties with grouping of components, systems, basic events, top events, and respective evaluations when interpreting the results.

For example, basic event name: xxx–yyy–zzzzz–ww can be formed from its parts as:

xxx System identification
yyy Subsystem identification
zzzzz Component identification
ww Failure mode identification

5.3.4 Fault Tree Construction

The fault tree construction is a step, where the fault tree model is developed graphically or where the Boolean equations which represent the fault tree are developed.

If the fault tree is developed graphically, the fault tree symbols are used. Figure 5.3 shows the fault tree symbols, although some others can be used in addition, such as EXCLUSIVE OR gate, PRIORITY AND gate, and INHIBIT gate.

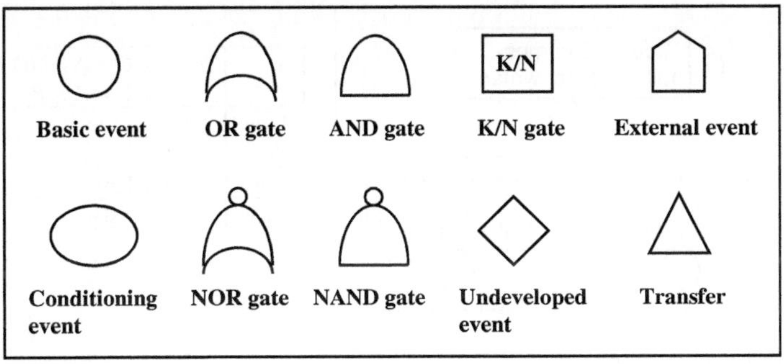

Fig. 5.3 Fault tree symbols

Fig. 5.4 Basic events examples

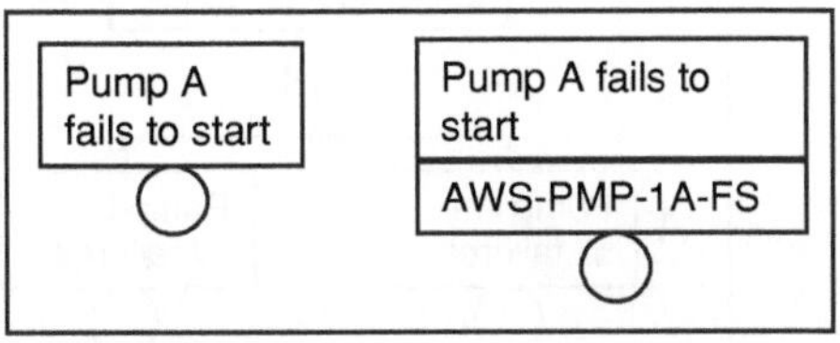

Fig. 5.5 Gates examples

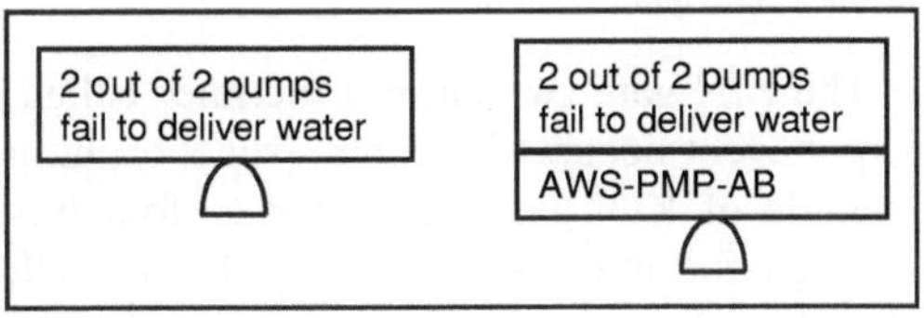

The first fault tree symbol is circle and represents the basic event. Figure 5.4 shows examples of basic events. The left one includes description and circle determining the basic event. The second one includes description, identification code and circle determining the basic event.

The fault tree construction starts with the top event, which can be mostly either OR gate or AND gate, or K/N gate. Gate can include description, identification code, and sign determining the logic of the gate. Figure 5.5 shows two variations of the gate, where the top event is represented by AND gate.

The AND gate indicates that the output event occurs if all of the input events occur at the same time. Figure 5.6 shows an example of AND gate connected to two basic events. The event *G* occurs if both events *A* and *B* occur, which is shown on the respective logic table.

The OR gate indicates that the output event occurs if any of the input events occur. Figure 5.7 shows an example of OR gate connected to two basic events. The event *G* occurs if any of both events *A* and *B* occurs, which is shown on the respective logic table.

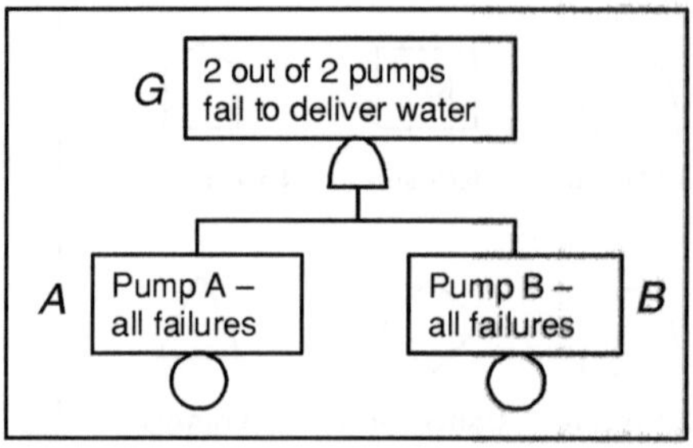

A	B	$G = A$ AND B $G = A \times B$
True	True	True
True	False	False
False	True	False
False	False	False

Fig. 5.6 AND gate example

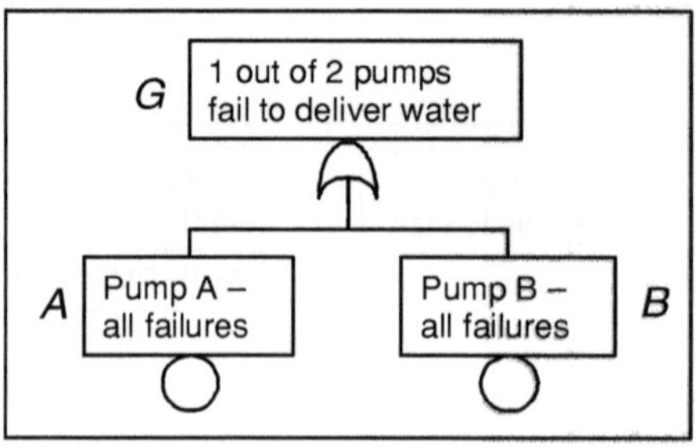

A	B	$G = A$ OR B $G = A + B$
True	True	True
True	False	True
False	True	True
False	False	False

Fig. 5.7 OR gate example

The K/N gate (which is sometimes called combination gate) indicates that the output event occurs if K input events occur at the same time. Figure 5.8 shows an example of K/N gate connected to four basic events. Three pumps failures are required for the system failure. From such top event with logic ¾, one can determine the success criteria of the system. Success criteria of the system would be a success of at least of 2 out of 4 pumps to deliver water.

The NOR gate indicates the negated OR gate.

The NAND gate indicates the negated AND gate.

The conditioning event represents any conditions or restrictions that apply to any logic gate.

The external event or house event is an event, which actually represents a logic switch. It mostly does not represent faults, but events such as phase change in a dynamic system or a change of mode of operation in a static system [5, 24, 25].

Figure 5.9 shows house event example under OR gate. The event $G1$ occurs if any of events $B1$ and $H1$ occur, which is shown on the respective logic table, which is the same as for the OR gate. The following feature is important: If $H1$ is true, the $G1$ is true no matter what is the state of event $B1$. If $H1$ is false, then the event $G1$ is of the same logic value as the event $B1$.

$$G_1 = H_1 + B_1; \quad \text{if } H_1 = \text{true} \Rightarrow G_1 = \text{true}; \quad \text{if } H_1 = \text{false} \Rightarrow G_1 = B_1.$$

Figure 5.10 shows house event example under AND gate. The event $G1$ occurs if both events $B1$ and $H1$ occur, which is shown on the respective logic table,

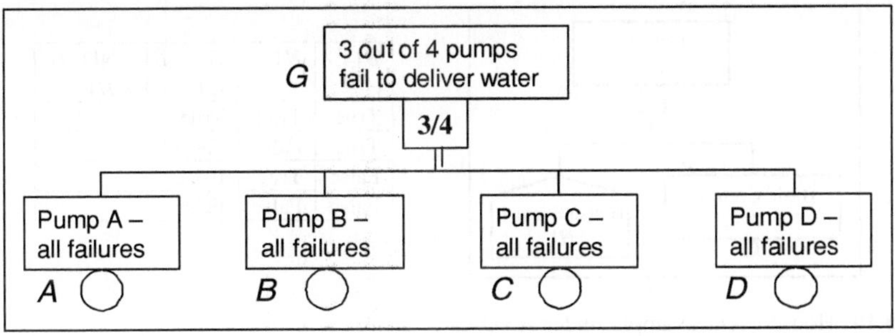

A	*B*	*C*	*D*	*G* = 3/4 (*A B C D*)
True	True	True	True	True
True	True	True	False	True
True	True	False	True	True
True	True	False	False	False
True	False	True	True	True
True	False	True	False	False
True	False	False	True	False
True	False	False	False	False
False	True	True	True	True
False	True	True	False	False
False	True	False	True	False
False	True	False	False	False
False	False	True	True	False
False	False	True	False	False
False	False	False	True	False
False	False	False	False	False

Fig. 5.8 K/N gate example

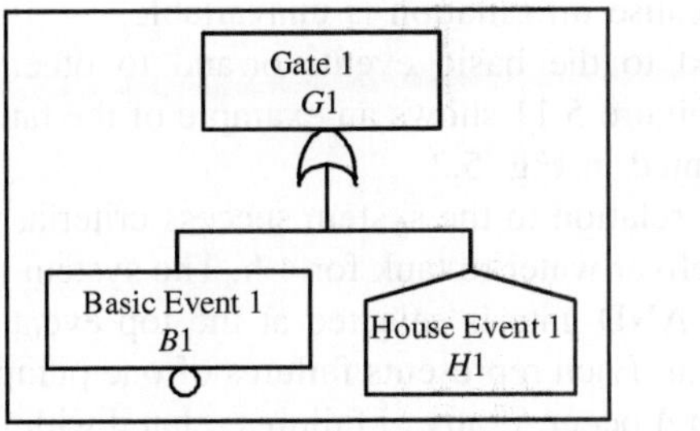

*B*1	*H*1	*G*1 = *B*1 OR *H*1 *G*1 = *B*1 + *H*1
True	True	True
True	False	True
False	True	True
False	False	False

Fig. 5.9 House event example under OR gate

which is the same as for the AND gate. The following feature is important: If $H1$ is false, the $G1$ is false no matter what is the state of event $B1$. If $H1$ is true, then the event $G1$ is of the same logic value as the event $B1$.

$$G_1 = H_1 \times B_1; \quad \text{if } H_1 = \text{false} \Rightarrow G_1 = \text{false}; \quad \text{if } H_1 = \text{true} \Rightarrow G_1 = B_1.$$

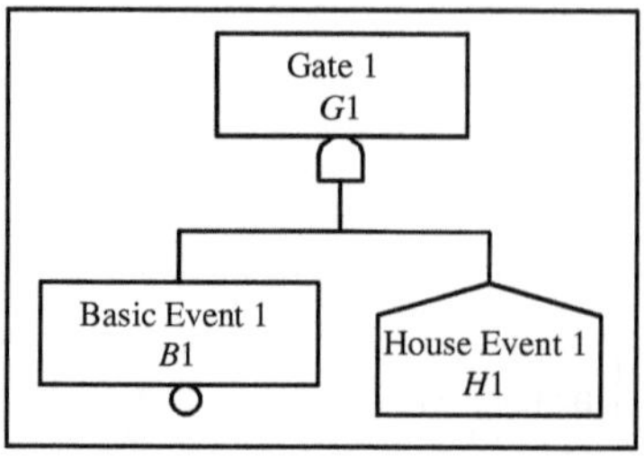

$B1$	$H1$	$G1 = B1$ AND $H1$ $G1 = B1 \times H1$
True	True	True
True	False	False
False	True	False
False	False	False

Fig. 5.10 House event example under AND gate

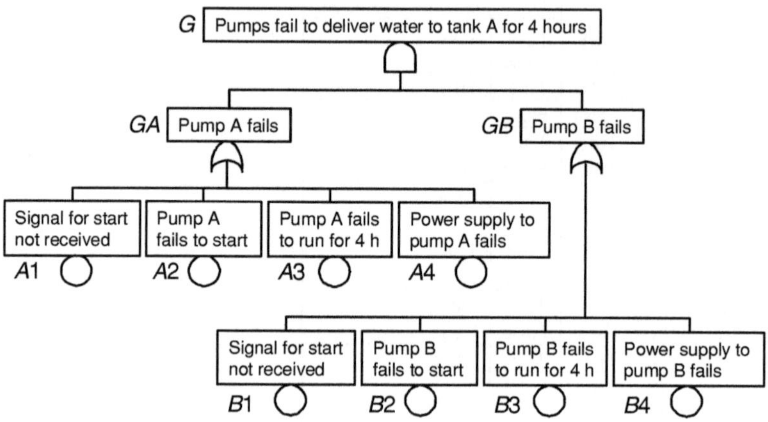

Fig. 5.11 Fault tree example of a simple example system

Undeveloped event is an event, which is not further developed either because it is of insufficient consequence or because information is unavailable.

The top event is then connected to the basic events or/and to other gates depending on the system structure. Figure 5.11 shows an example of the fault tree for a simple system, which is presented in Fig. 5.2.

The top event is determined with relation to the system success criteria, which require that one out of two pumps deliver water to tank for 4 h. The system failure would occur if both pumps fail, so AND gate is selected at the top event. Both input gates to the top event are similar. Each represents failures of one pump. Left gate connected with failures of pump *A* occur, if any of failures related with failure of pump *A* occur, so OR gate is selected, which is linked to four basic events. Every basic event represents such failure of such equipment, which leads to failure of pump *A*.

The graphic representation of the fault tree can be identically written as the set of logic or Boolean equations, which define relations between events in the fault tree. The Boolean equations for the fault tree in Fig. 5.11 are the following:

$G = GA \times GB$, which links the top event *G* through AND gate to the gates *GA* and *GB*; both *GA* and *GB* should occur for the occurrence of *G*;

$GA = A1 + A2 + A3 + A4$, which links gate GA through OR gate with basic events $A1$, $A2$, $A3$, and $A4$. Any of the events $A1$ OR $A2$ OR $A3$ OR $A4$ occur for the occurrence of GA;

$GB = B1 + B2 + B3 + B4$, which links gate GB through OR gate with basic events $B1$, $B2$, $B3$, and $B4$. Any of the events $B1$ OR $B2$ OR $B3$ OR $B4$ occur for the occurrence of GB.

5.3.5 Qualitative Fault Tree Evaluation

Qualitative fault tree evaluation is the process of finding the combinations of basic events, which, if they occur, cause the top event occurrence.

If the fault tree is written in the form of Boolean equations, those need to be combined into one with applying the rules of Boolean algebra to obtain the equation for top event, which consists of sum of products of basic events.

If the fault tree is developed in its graphical form, the Boolean equations need to be written first based on the logic of the gates and their inputs. And then, the rules of the Boolean algebra are applied to obtain the equation for top event, which consists of sum of products of basic events.

The sum of products of basic events identifies the minimal cut sets. The minimal cut sets are the combinations of components failures, which fail the system. In other words, the minimal cut sets are combinations of the smallest number of basic events, which if occur simultaneously, may lead to the top event.

When the sum of products of basic events is expressed from Boolean equations of the fault tree, each element of the sum includes the product of a certain number of basic events. Those basic events together represent a minimal cut set.

A minimal cut set can include only one basic event. This is then single minimal cut set. Single minimal cut set means the single-component failure can fail the system, which is under consideration. A minimal cut set can include two basic events. This is then double minimal cut set. Double minimal cut set means that two specific component failures can fail the system. A minimal cut set can include three basic events. This is then triple minimal cut set. Triple minimal cut set means that three specific component failures can fail the system. A minimal cut set can include four basic events. This is then quadruple minimal cut set. Quadruple minimal cut set means that four specific component failures can fail the system. A minimal cut set can include any larger number of basic events. The larger this number is, the less probable system failure would be.

For the simplest example of the fault tree, which is presented in Fig. 5.6, only one short equation is representing the fault tree and at the same time it shows the qualitative result, which shows the sum of products of events. In this particular event, it is only one set of products: $G = A \times B$.

This equation shows that only one minimal cut set exists and it includes two events A and B. If both fail, would cause system to fail. This particular case includes one double minimal cut set.

Table 5.1 Boolean algebra

Commutative law	$X \cap Y = Y \cap X$; $X \times Y = Y \times X$; $X \cup Y = Y \cup X$; $X + Y = Y + X$
Associative law	$X \cap (Y \cap Z) = (X \cap Y) \cap Z$; $X \times (Y \times Z) = (X \times Y) \times Z$ $X \cup (Y \cup Z) = (X \cup Y) \cup Z$; $X + (Y + Z) = (X + Y) + Z$
Distributive law	$X \cap (Y \cup Z) = (X \cap Y) \cup (X \cap Z)$; $X \times (Y + Z) = X \times Y + X \times Z$ $X \cup (Y \cap Z) = (X \cup Y) \cap (X \cup Z)$; $X + Y \times Z = (X + Y) \times (X + Z)$
Idempotent law	$X \cap X = X$; $X \times X = X$; $X \cup X = X$; $X + X = X$
Law of absorption	$X \cap (X \cup Y) = X$; $X \times (X + Y) = X$; $X \cup (X \cap Y) = X$; $X + X \times Y = X$
Complementation	$X \cap X' = \varnothing$; $X \times X' = \varnothing$ ($\varnothing$ means empty set) $X \cup X' = \Omega = I$; $X + X' = \Omega = I$ (Ω or I means the universal set) $(X')' = X$ $(X')' = X$ (X' means negation of X)
de Morgan rules	$(X \cap Y)' = X' \cup Y'$; $(X \times Y)' = X' + Y'$; $(X \cup Y)' = X' \cap Y'$; $(X + Y)' = X' \times Y'$
Operations with $\varnothing$ and Ω	$\varnothing \cap X = \varnothing$; $\varnothing \times X = \varnothing$; $\varnothing \cup X = X$; $\varnothing + X = X$; $\Omega \cap X = X$; $\Omega \times X = X$; $\Omega \cup X = \Omega$; $\Omega + X = \Omega$; $\varnothing' = \Omega$; $\varnothing' = \Omega$; $\Omega' = \varnothing$; $\Omega' = \varnothing$.
Other relationships	$X \cup (X' \cap Y) = X \cup Y$; $X + X' \times Y = X + Y$; $X' \cap (X \cup Y') = X' \cap Y' = (X \cup Y)'$; $X' \times (X + Y') = X' \times Y' = (X + Y)'$

For the simplest example of the fault tree, which is presented in Fig. 5.7, only one short equation is representing the fault tree and at the same time it shows the qualitative result, which shows the sum of products of events. In this particular event, it is only two set of products: $G = A + B$.

This equation shows that only two minimal cut set exist and they both include one event. The first minimal cut set includes the event A and the second minimal cut set includes the event B. If event A occurs, this would cause the system to fail. Or if event B occurs, this would cause the system to fail. This particular case includes two single minimal cut sets.

The rules of Boolean algebra, which are needed for evaluation of more complex fault trees, are summarized in Table 5.1.

For the fault tree, which is presented in Fig. 5.11, the set of equations is the following:

$$G = GA \times GB \tag{5.1}$$

$$GA = A1 + A2 + A3 + A4 \tag{5.2}$$

$$GB = B1 + B2 + B3 + B4 \tag{5.3}$$

If the second and the third equation are inserted into the first equation, only one equation is obtained:

$$G = (A1 + A2 + A3 + A4) \times (B1 + B2 + B3 + B4) \tag{5.4}$$

Evaluation of the product gives the final equation, which consists of 16 minimal cut sets, which are all double minimal cut sets in this particular example, and which means that each include two basic events:

$$
\begin{aligned}
G = {} & A1 \times B1 + A2 \times B1 + A3 \times B1 + A4 \times B1 + A1 \times B2 + A2 \times B2 + A3 \\
& \times B2 + A4 \times B2 + A1 \times B3 + A2 \times B3 + A3 \times B3 + A4 \times B3 + A1 \times B4 \\
& + A2 \times B4 + A3 \times B4 + A4 \times B4
\end{aligned}
\quad (5.5)
$$

If events both $A1$ and $B1$ occur, the system fails. Similarly, this is valid for any other double minimal cut sets, e.g., event $A2$ occurs and event $B1$ occurs, all together 16 double minimal cut sets.

In general, the equations for representing the minimal cut sets as the result of the qualitative fault tree analysis are joined into the following equation.

$$G = \sum_{i=1}^{I} \prod_{j=1}^{J} B_j$$

where G is the top event, B_j is the basic event j, J is the number of basic events in a particular minimal cut set, and I is the number of minimal cut sets.

5.3.6 Preparation of the Probabilistic Failure Database

Preparation of the probabilistic failure database and connection of the basic events of the fault tree with probabilistic failure data is a prerequisite for quantitative evaluation of the fault tree analysis [26–30]. It is necessary to assess the failure probability of each basic event or its unavailability.

Preparation of the probabilistic failure database includes the following activities:

- Selection of probabilistic model
- Preparation of probabilistic failure database
- Link of a probabilistic model with the appropriate data from database

5.3.6.1 Selection of Probabilistic Model

Probabilistic models are developed to assess component failure probabilities [31]. The variety of probabilistic models is needed, because of the different nature of considered components, their functioning role in the fulfilling the success criteria of the system operation, and their failure modes.

For example, assessment of failure probability of a switch, which has to open or close, is surely different than the failure probability of a running pump, which has to run for a certain time interval.

Or, for example, the failure probability is surely different for a pump, which is running and has to be running for 4 h than for a pump, which has to start first and then run for 24 h. Or, for example, the failure probability is surely different for valve that has to open for accomplishment of its mission compared with a valve that has to remain open for accomplishment of its mission.

The simplest probabilistic model is developed for components that have to operate on demand. Their failure probability can be assessed through the ratio of number of unsuccessful operations versus the number of all operations.

$$p = \frac{n_f}{n} \tag{5.6}$$

where p is the component failure probability, n_f is the number of unsuccessful operations, and n is the number of all operations.

Or the model assumes that the number of failures has a binomial distribution. On each demand, the outcome is a failure with some probability p, and a success with probability $1 - p$. This probability p is the same for all demands. Occurrences of failures for different demands are statistically independent, which means the probability of a failure on one demand is not affected by what happens on other demands. The total number of failures and the total number of demands are observed as a minimum. Under these assumptions, the random number of failures in some fixed number of demands has a binomial (n, p) distribution.

$$p = p(k) = P(X = k) = \binom{n}{k} p^k (1 - p)^{n-k} \tag{5.7}$$

where X is the random number of failures, and n is the fixed number of demands.

If the component is assumed to transform to the failed state while the system is in standby, the transition occurs at a random time with a constant transition rate. The latent failed condition ensures that the system fails at the next demand, but the condition is not discovered until the next test, the inspection or actual demand. The probability that the system is failed when observed at time t is calculated knowing the failure rate and the equation.

$$p = 1 - e^{-\lambda t} \tag{5.8}$$

where λ is the failure rate and t is time.

The failure to run during mission requires different approach, actually two approaches, which differ, if the failure is repairable or it is not repairable.

If the failure is repairable, it does not cause directly the mission failure. The parameters, mean time to repair or the mean time to restore, are important in addition to the failure rate of component in operation λ_0 for calculation of the component failure probability.

$$p = \frac{\lambda_o T_r}{1 + \lambda_o T_r} \tag{5.9}$$

where T_r is the mean time to restore (or mean time to repair) and λ_o is the failure rate of component in operation.

The other option of the equation for repairable component is equation combining failure rate and repair rate of the component.

$$p = \frac{\lambda}{\lambda + \mu} \tag{5.10}$$

where μ is the repair rate.

If the failure is not repairable, it causes the mission failure. Specially, for missions of a short duration, it is difficult to expect that the failure can be a repairable, so the nonrepairable model is used considering parameters standby failure rate and lifetime.

$$p = 1 - e^{-\lambda_s T_{pl}} \tag{5.11}$$

where T_{pl} is the lifetime and λ_s is the standby failure rate.

The testing and maintenance contribute to unavailability of standby components. The unavailability of testing is larger if the testing time is of longer duration or if the test interval is shorter.

As the definition of unavailability says, it is the probability that certain event occurs under certain conditions in determined time interval, this unavailability of testing can be added to failure probability of the component to combine both to represent the component quantification in more details.

$$Q_t = \frac{T_t}{T_i} \tag{5.12}$$

where T_t is the testing time and T_i is the test interval.

Similarly, it is for scheduled maintenance, where the frequency of scheduled maintenance per test interval, maintenance duration time and maintenance interval are the parameters of interest.

$$Q_{sm} = f_m \frac{T_{ma}}{T_{mi}} \tag{5.13}$$

where f_m is the frequency of scheduled maintenance per test interval, T_{ma} is the maintenance duration time, and T_{mi} is the maintenance interval.

Similarly, it is for unscheduled repair, where the frequency of unscheduled repair per test interval, mean time to repair and test interval are the parameters of interest.

$$Q_{ur} = f_r \frac{T_r}{T_i} \tag{5.14}$$

where f_r is the frequency of unscheduled repair per test interval.

The combinations of certain probabilistic models are possible, specially, if the testing or maintenance is considered for standby equipment in addition to the selected probabilistic model related to failure mode.

Several other probabilistic models exist in theory and each is a function of one or more parameters of interest. The parameters of the probabilistic model have to be obtained from the database.

5.3.6.2 Preparation of Probabilistic Failure Database

The collection and classification of the probabilistic data are a time-demanding and complex process [29]. The data for preparation of probabilistic failure database can be derived from different sources, including other databases, description documents of the plant or facility, reports, industry experience, experience of similar plants or facilities, and experience of the particular plant or facility.

The collection and classification of the probabilistic data have to be oriented to obtain all required parameters that support the evaluation of the probabilistic models. The parameters of interest include the following: failure rate, failure probability on demand, mission time, mean time to repair, mean time to failure, testing interval, testing duration time, maintenance interval, maintenance duration time, frequency of scheduled maintenance per test interval, and frequency of unscheduled repair per test interval.

The parameters of interest are selected to cover the failure modes that are defined in the basic events of the fault tree. Those failure modes include failure to change position, failure to remain in position, failure to close, failure to open, failure to function, short to ground, short circuit, open circuit, plugging, rupture, spurious function, failure to run, failure to start, leaking, and overheating.

The parameters of the interest are selected to cover all components that are modeled in the fault trees, such as piping, heat exchangers, pumps, valves, strainers, filters, transformers, relays, motors, conductors, batteries and chargers, circuit breakers, disconnect switches, power lines, power generators, sensors, instrumentation channels, transmitters, signal-conditioning systems, switches, and other mechanical or electrical or instrumentation equipment.

Examples of databases are in the references [26–28]. Not only the mean value of the required parameter is contained in the database but also the parameters of probability distribution can be of interest to consider the uncertainty of the data [30].

5.3.6.3 Link of a Probabilistic Model With the Appropriate Data from Database

When the probabilistic model for a specific basic event is selected, the related parameters that support this probabilistic model are linked from database to specific basic event. The selection of the data for a probabilistic model enables quantification of the failure probability of respective basic event. Figure 5.12

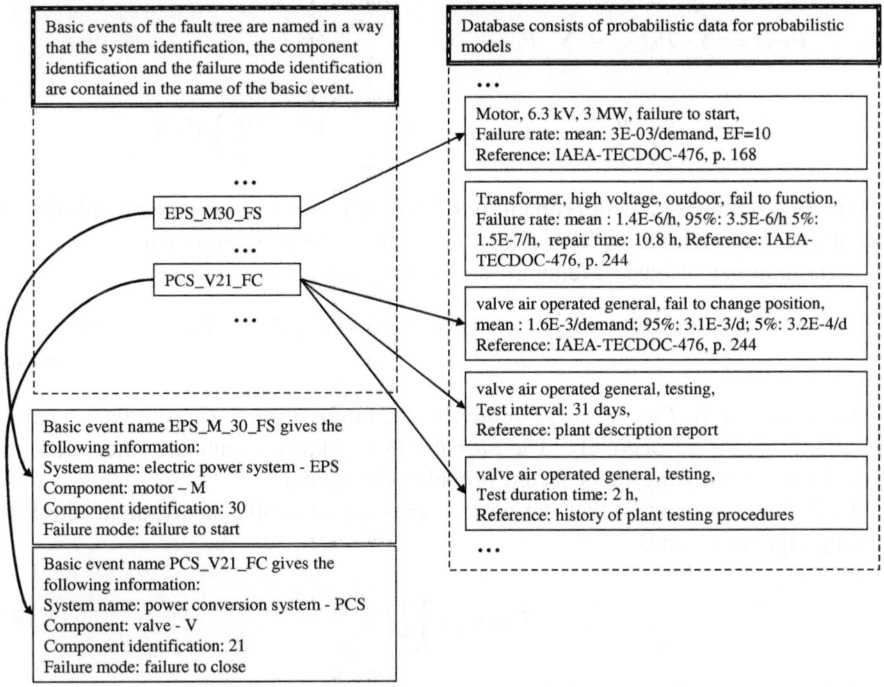

Fig. 5.12 Link of example basic events with selected parameters from database

shows two example basic events and their link with parameters from the database. One basic event is linked to only one parameter as the basic event description is related only to failure to start. The second basic event is related to more parameters as the basic event description of the second basic event requires probabilistic model, which requires three parameters.

5.3.7 Quantitative Fault Tree Evaluation

When the probabilistic data are assigned to all basic events of the fault tree, the basic event probabilities can be calculated for all basic events. When the basic event probabilities are known, they propagate through the fault tree up to the top event regarding the fault tree configuration.

The alternative way of evaluation of the fault tree is the use of generated minimal cut sets from qualitative fault tree analysis and this way is mostly used in the computer codes dealing with the fault tree evaluation. The quantitative result of the fault tree evaluation is the top event probability, which is a representation of the system failure probability. The equation is the following.

$$P_{\mathrm{TOP}} = \sum_{i=1}^{n} P_{\mathrm{MCS}_i} - \sum_{i<j} P_{\mathrm{MCS}_i \cap \mathrm{MCS}_j} + \sum_{i<j<k} P_{\mathrm{MCS}_i \cap \mathrm{MCS}_j \cap \mathrm{MCS}_k} - \cdots + (-1)^{m-1} P \bigcap_{i=1}^{m} \mathrm{MCS}_i \tag{5.15}$$

where P_{TOP} is the top event probability of the fault tree, $P_{\mathrm{MCS}\,i}$ is the probability of occurrence of minimal cut set i (MCS_i), n is the number of minimal cut sets, and m is the number of basic events in the largest minimal cut set.

$$P_{\mathrm{MCS}_i} = P_{B1} \times P_{B2}|P_{B1} \times P_{B3}|P_{B1} \cap P_{B2} \times \cdots \times P_{Bm}|P_{B1} \cap P_{B2} \cap \cdots \cap P_{Bm-1} \tag{5.16}$$

Parameters $P_{B1}, P_{B2}, \ldots, P_{Bn}$, represent the failure probabilities of basic events $B1, B2, \ldots, Bn$, respectively. The probabilities of basic events are calculated for each basic event using the selected probabilistic model.

Under the assumption that the basic events are mutually independent, the following equation stands.

$$P_{\mathrm{MCS}_i} = \prod_{j=1}^{m} P_{Bj} \tag{5.17}$$

Failure probability of each basic event is a function of respective parameters from the selected probabilistic model. The general equation for all probabilistic models is represented by the following equation, where the failure probability of the respective basic event is a function of only certain parameters, which are connected with the respective probabilistic model.

$$P_{Bj} = P_{Bj}(\lambda_{Bj}, p_{Bj}, T_{iBj}, T_{tBj}, T_{mBj}, T_{rBj}, \eta_{Bj}) \tag{5.18}$$

where λ_{Bj} is the failure rate of equipment modeled in basic event Bj, p_{Bj} is the probability per demand of equipment modeled in basic event Bj, T_{iBj} is the test interval of equipment modeled in basic event Bj, T_{tBj} is the test duration of equipment modeled in basic event Bj, T_{mBj} is the mission time of equipment modeled in basic event Bj, T_{rBj} is the repair time of equipment modeled in basic event Bj, and μ_{Bj} is the repair rate of equipment modeled in basic event Bj.

As the equation representing the evaluation of the fault tree from the use of minimal cut sets may be too complex for larger fault trees, its simplified version may be used, which considers only a selected number of summands, e.g., three, for example:

$$P_{\mathrm{TOP}} = \sum_{i=1}^{n} P_{\mathrm{MCS}_i} - \sum_{i<j} P_{\mathrm{MCS}_i \cap \mathrm{MCS}_j} + \sum_{i<j<k} P_{\mathrm{MCS}_i \cap \mathrm{MCS}_j \cap \mathrm{MCS}_k} \tag{5.19}$$

Sometimes, a consideration of only the first summand is sufficient, specially, if we bear in mind that the conservative value is given with such a simplification.

Such approximation is called the first-order approximation. For P_{MCS_i} less than 0.1, the approximate results stay in 10% of accuracy in the conservative side [4, 5, 14].

$$P_{\mathrm{TOP}} = \sum_{i=1}^{n} P_{\mathrm{MCS}_i} \tag{5.20}$$

If for the example system from Fig. 5.11 the failure probabilities of all components equal to 1E-2, then the quantification of the top event probability gives the system failure probability of 0.148.

$$\begin{aligned} P_{\mathrm{TOP}} &= \sum_{i=1}^{16} 10^{-2} \times 10^{-2} - \sum_{1}^{15+14+\cdots+1} 10^{-2} \times 10^{-2} \times 10^{-2} \times 10^{-2} \\ &= 0.16 - 0.012 = 0.148 \end{aligned} \tag{5.21}$$

The quantitative results include the importance factors in addition [32]. The importance factors include Fussel–Vesely importance, risk achievement worth, risk reduction worth, Birnbaum importance, criticality importance, and differential importance measure. The various importance measures are based on slightly different interpretations of the concept of component importance. In general, the importance of a component within a system depends on the location of the component in the system, onthe reliability of the component, and the reliability of the system. The importance measures can be quantified for each of the basic events, if the interest about specific component is important, or for the groups of basic events, if the group of the components is important.

5.3.7.1 Fussel–Vesely Importance

Fussel–Vesely importance shows the contribution of the event to the top event probability. It provides a numerical significance of all the events as parts of the fault tree and allows them to be prioritized. It is calculated according to the following equation.

$$\mathrm{FV}_k = 1 - \frac{P_{\mathrm{TOP}}(P_k = 0)}{P_{\mathrm{TOP}}} = 1 - \frac{1}{\mathrm{RRW}_k} \tag{5.22}$$

where FV_k is the Fussel–Vesely importance for component modeled in basic event k, P_{TOP} is the top event probability, $P_{\mathrm{TOP}}(P_k = 0)$ is the top event probability when failure probability of component modeled in basic event k is set to 0, and RRW_k is the risk reduction worth for component k.

For the example system from Fig. 5.11, the Fussel–Vesely is calculated for the event $A1$ knowing that $P_{\mathrm{TOP}}(P_k = 0)$ is calculated from 12 minimal cut sets without 4 minimal cut sets, which equal 0, because they contain $P_{A1} = 0$.

A, B, C, D ... Basic events
TOP ... Top event
$G1, G2$ ... Gates
Logic equations (ekvivalence to fault tree):
TOP = $A + G1$
$G1 = B \times G2$
$G2 = C + D$
Minimal cut sets: TOP = $A + BC + BD$
Basic event failure probabilities: P_A, P_B, P_C, P_D
Top event probability: $P_{\mathrm{TOP}} = P_A + P_B \times P_C + P_B \times P_D$

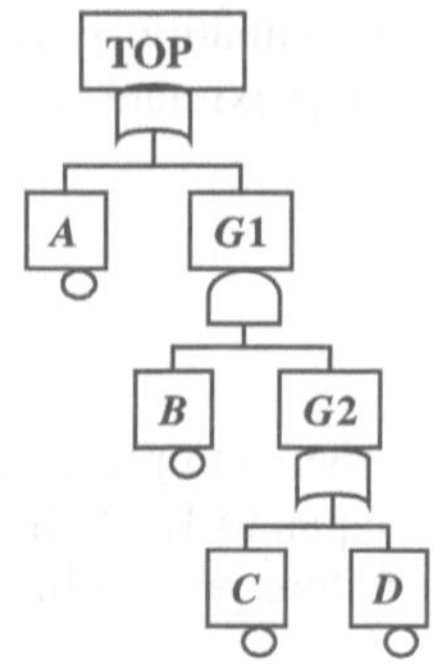

Fig. 5.13 Simple fault tree example for calculation of basic events importance

$$P_{\mathrm{TOP}}(P_k = 0) = \sum_{i=1}^{12} 10^{-2} \times 10^{-2} - \sum_{1}^{11+10+\cdots+1} 10^{-2} \times 10^{-2} \times 10^{-2} \times 10^{-2}$$
$$= 0.12 - 0.0066 = 0.1134 \tag{5.23}$$

$$\mathrm{FV}_k = 1 - \frac{P_{\mathrm{TOP}}(P_k = 0)}{P_{\mathrm{TOP}}} = 1 - \frac{0.1134}{0.148} = 0.234 \tag{5.24}$$

Figure 5.13 shows a better example fault tree of an example system, which is not completely symmetrical. The fault tree with the top event named TOP is presented in figure and with Boolean equations. The basic events, gates, and top event are represented by their names. The minimal cut sets are shown, which are the results of the qualitative analysis. The basic event failure probabilities give the component failure probabilities. The equation for the top event calculation is derived based on minimal cut sets, which represents the system failure probability through the calculation of the fault tree top event probability P_{TOP} with the first-order approximation.

The following equations represent the calculations of the Fussel–Vesely importance for all components of the system presented in Fig. 5.13.

$$\mathrm{FV}_A = 1 - \frac{P_{\mathrm{TOP}}(P_A = 0)}{P_A + P_B P_C + P_B P_D} = 1 - \frac{P_B P_C + P_B P_D}{P_A + P_B P_C + P_B P_D}$$
$$= \frac{P_A}{P_A + P_B P_C + P_B P_D} \tag{5.25}$$

$$\mathrm{FV}_B = 1 - \frac{P_{\mathrm{TOP}}(P_B = 0)}{P_A + P_B P_C + P_B P_D} = 1 - \frac{P_A}{P_A + P_B P_C + P_B P_D}$$
$$= \frac{P_B P_C + P_B P_D}{P_A + P_B P_C + P_B P_D} \tag{5.26}$$

$$\mathrm{FV}_C = 1 - \frac{P_{\mathrm{TOP}}(P_C = 0)}{P_A + P_B P_C + P_B P_D} = 1 - \frac{P_A + P_B P_D}{P_A + P_B P_C + P_B P_D}$$
$$= \frac{P_B P_C}{P_A + P_B P_C + P_B P_D} \tag{5.27}$$

$$\mathrm{FV}_D = 1 - \frac{P_{\mathrm{TOP}}(P_D = 0)}{P_A + P_B P_C + P_B P_D} = 1 - \frac{P_A + P_B P_C}{P_A + P_B P_C + P_B P_D}$$
$$= \frac{P_B P_D}{P_A + P_B P_C + P_B P_D} \tag{5.28}$$

The following equation represents the Fussel–Vesely importance measure for a group of components: component modeled in basic event C and component modeled in basic event D, for the example system presented in Fig. 5.13.

$$\mathrm{FV}_{C,D} = 1 - \frac{P_{\mathrm{TOP}}(P_C = 0, P_D = 0)}{P_A + P_B P_C + P_B P_D} = 1 - \frac{P_A}{P_A + P_B P_C + P_B P_D}$$
$$= \frac{P_B P_C + P_B P_D}{P_A + P_B P_C + P_B P_D} \tag{5.29}$$

5.3.7.2 Risk Achievement Worth

The risk achievement worth for a basic event shows the increase in the probability of the top event that would be obtained if the lower-level event, e.g., the failure of a component modeled in the basic event, would occur. The risk achievement worth shows basic events where prevention activities should be focused to assure failures do not occur. The large risk achievement worth identifies the basic events, which contain components, which are worth to maintain very well in order that the overall risk is not significantly increased. The risk achievement worth is known also by the term risk increase factor.

$$\mathrm{RAW}_k = \frac{P_{\mathrm{TOP}}(P_k = 1)}{P_{\mathrm{TOP}}} \tag{5.30}$$

where RAW_k is the risk achievement worth for component modeled in basic event k, $P_{\mathrm{TOP}}(P_k = 1)$ is the top event probability when failure probability of component modeled in basic event k is set to 1, and P_{TOP} is the top event probability.

The following equations represent the calculations of the risk achievement worth for all components of the system for the example fault tree of an example system presented in Fig. 5.13.

$$\mathrm{RAW}_A = \frac{P_{\mathrm{TOP}}(P_A = 1)}{P_A + P_B P_C + P_B P_D} = \frac{1 + P_B P_C + P_B P_D}{P_A + P_B P_C + P_B P_D} \tag{5.31}$$

$$\mathrm{RAW}_B = \frac{P_{\mathrm{TOP}}(P_B = 1)}{P_A + P_B P_C + P_B P_D} = \frac{P_A + P_C + P_D}{P_A + P_B P_C + P_B P_D} \tag{5.32}$$

$$\mathrm{RAW}_C = \frac{P_{\mathrm{TOP}}(P_C = 1)}{P_A + P_B P_C + P_B P_D} = \frac{P_A + P_B + P_B P_D}{P_A + P_B P_C + P_B P_D} \tag{5.33}$$

$$\mathrm{RAW}_D = \frac{P_{\mathrm{TOP}}(P_D = 1)}{P_A + P_B P_C + P_B P_D} = \frac{P_A + P_B P_C + P_B}{P_A + P_B P_C + P_B P_D} \tag{5.34}$$

5.3.7.3 Risk Reduction Worth

The risk reduction worth for a basic event shows the decrease in the probability of the top event that would be obtained if the lower-level event, e.g., the failure of a component modeled in the basic event, did not occur. The large risk reduction worth implies that the risk of the respective basic event is worth to decrease in order that the overall risk is significantly decreased. The large risk reduction worth implies that the risk of the respective components modeled in their respective basic events is worth to decrease in order that the overall risk is significantly decreased. The risk reduction worth identifies components that are candidates for redundancy. The risk reduction worth is known also by the term risk decrease factor.

$$\mathrm{RRW}_k = \frac{P_{\mathrm{TOP}}}{P_{\mathrm{TOP}}(P_k = 0)} \tag{5.35}$$

where P_{TOP} is the top event probability, $P_{\mathrm{TOP}}(P_k = 0)$ is the top event probability when failure probability of component modeled in basic event k is set to 0, and RRW_k is the risk reduction worth for component modeled in basic event k.

The following equations represent the calculations of the risk reduction worth for all components of the system for the example fault tree of an example system presented in Fig. 5.13.

$$\mathrm{RRW}_A = \frac{P_A + P_B P_C + P_B P_D}{P_{\mathrm{TOP}}(P_A = 0)} = \frac{P_A + P_B P_C + P_B P_D}{P_B P_C + P_B P_D} \tag{5.36}$$

$$\mathrm{RRW}_B = \frac{P_A + P_B P_C + P_B P_D}{P_{\mathrm{TOP}}(P_B = 0)} = \frac{P_A + P_B P_C + P_B P_D}{P_A} \tag{5.37}$$

$$\mathrm{RRW}_C = \frac{P_A + P_B P_C + P_B P_D}{P_{\mathrm{TOP}}(P_C = 0)} = \frac{P_A + P_B P_C + P_B P_D}{P_A + P_B P_D} \tag{5.38}$$

$$\mathrm{RRW}_D = \frac{P_A + P_B P_C + P_B P_D}{P_{\mathrm{TOP}}(P_D = 0)} = \frac{P_A + P_B P_C + P_B P_D}{P_A + P_B P_C} \tag{5.39}$$

5.3.7.4 Birnbaum Importance

The Birnbaum importance represents the rate of change in the top event probability as a result of the change in the probability of a given event. The Birnbaum importance can be calculated by first calculating the top event probability with the probability of the given event set to 1 and then subtracting the top event probability with the probability of the given event set to 0.

$$B_k = P_{\mathrm{TOP}}(P_k = 1) - P_{\mathrm{TOP}}(P_k = 0) \tag{5.40}$$

where $P_{\mathrm{TOP}}(P_k = 1)$ is the top event probability when failure probability of component modeled in basic event k is set to 1, and $P_{\mathrm{TOP}}(P_k = 0)$ is the top event probability when failure probability of component modeled in basic event k is set to 0.

The following equations represent the calculations of the Birnbaum importance for all components of the system for the example fault tree of an example system presented in Fig. 5.13.

$$\begin{aligned} B_A &= P_{\mathrm{TOP}}(P_A = 1) - P_{\mathrm{TOP}}(P_A = 0) \\ &= 1 + P_B \times P_C + P_B \times P_D - P_B \times P_C - P_B \times P_D = 1 \end{aligned} \tag{5.41}$$

$$B_B = P_{\mathrm{TOP}}(P_B = 1) - P_{\mathrm{TOP}}(P_B = 0) = P_A + P_C + P_D - P_A = P_C + P_D \tag{5.42}$$

$$\begin{aligned} B_C &= P_{\mathrm{TOP}}(P_C = 1) - P_{\mathrm{TOP}}(P_C = 0) = P_A + P_B + P_B \times P_D - P_A - P_B \times P_D \\ &= P_B \end{aligned} \tag{5.43}$$

$$\begin{aligned} B_D &= P_{\mathrm{TOP}}(P_D = 1) - P_{\mathrm{TOP}}(P_D = 0) = P_A + P_B \times P_C + P_B - P_A - P_B \times P_C \\ &= P_B \end{aligned} \tag{5.44}$$

5.3.7.5 Criticality Importance

The criticality importance represents the rate of change in the top event probability as a result of the change in the probability of a given event relatively to the top event probability and failure probability of component modeled in basic event k.

$$C_k = \frac{P_{\mathrm{TOP}}(P_k = 1) - P_{\mathrm{TOP}}(P_k = 0)}{P_{\mathrm{TOP}}} P_k \tag{5.45}$$

where $P_{\mathrm{TOP}}(P_k = 1)$ is the top event probability when failure probability of component modeled in basic event k is set to 1, $P_{\mathrm{TOP}}(P_k = 0)$ is the top event probability when failure probability of component modeled in basic event k is set to 0, P_{TOP} is the top event probability, and P_k is the failure probability of component k.

5.3.7.6 Differential Importance Measure

Differential importance measure is in more details presented in reference [33].

5.3.8 Interpretation of the Fault Tree Analysis Results

The interpretation of the fault tree analysis results is a phase where the qualitative and quantitative results of the fault tree analysis are considered together with the assumptions and limitations of the analysis, with the boundary conditions of the analysis and the resolution of modeling. It is a phase where the sensitivity and uncertainty of the results are evaluated before the final conclusions, of what the results show, are made.

5.4 Applications of the Fault Tree Analysis

Applications of the fault tree analysis are numerous. Only a small number of selected applications are mentioned:

- Reliability studies of safety systems in nuclear and air and space industry [5, 13, 16, 21]
- Optimization of preventive maintenance in nuclear power plants
- Vulnerability studies [34, 35]
- Safety software quality improvement [36–38]

The methods for optimization of preventive maintenance in a large extent consider the standby safety equipment, because the majority of the safety systems in nuclear power plants are in standby. The methods include:

- Optimization of surveillance-testing intervals [39–43]
- Optimization of scheduling of testing and maintenance activities [42, 44]
- Optimization of testing strategies [45, 46]
- Optimization of allowed outage times [46, 47]

Common to optimization ideas and their implementations is a fact that the problems are becoming larger and larger, because the new methods and better computers in present time allow modeling of processes and properties, which were not considered in the past.

The bottleneck of maintenance optimization methods and their applications lays in difficulty of appropriate modeling of all positive and all negative aspects of maintenance in the probabilistic models, which are consequently used in optimization methods.

The number of papers about fault tree analysis in scientific literature shows that its importance is increasing through the years [48–80].

References

1. ASME RA-S-2002 (2002) Standard for probabilistic risk assessment for nuclear power plant applications, addendum (2005). ASME
2. RA-S-2008 (2008) Standard for level 1/large early release frequency probabilistic risk assessment for nuclear power plant applications. ASME
3. IEC 61025 (2006) Fault tree analysis (FTA). IEC
4. Roberts NH, Vesely WE, Haasl D, Goldberg FF (1981) Fault tree handbook, NUREG-0492. NRC, Washington
5. Vesely W, Dugan J, Fragola J et al (2002) Fault tree handbook with aerospace applications. National Aeronautics and Space Administration
6. Kumamoto H, Henley EJ (1996) Probabilistic risk assessment and management for engineers and scientists. IEEE, New York
7. Villemeur A (1992) Reliability, availability, maintainability and safety assessment: methods and techniques. Wiley, New York
8. No 50-P-8 (1995) Procedures for conduction probabilistic safety assessments of nuclear power plants (level 2), Safety Series. IAEA
9. No 50-P-12 (1996) Procedures for conduction probabilistic safety assessments of nuclear power plants (level 3), Safety Series. IAEA
10. No 50-P-12 (1996) Procedures for conduction probabilistic safety assessments of nuclear power plants (level 3), Safety Series. IAEA
11. Probabilistic risk assessment procedures guide (1982) NUREG/CR-2300, NRC
12. Probabilistic safety analysis procedures guide (1985) NUREG/CR-2815, NRC
13. PRA NASA Guide (2002) Probabilistic risk assessment procedures guide for NASA managers and practitioners. NASA
14. Čepin M (2005) Analysis of truncation limit in probabilistic safety assessment. Rel Eng Syst Saf 87(3):395–403
15. Interim reliability evaluation program procedures guide (1983) NUREG/CR-2728, NRC
16. Analysis of core damage frequency (1990) NUREG/CR-4550, NRC
17. WASH-1400 (1975) Reactor safety study: an assessment of accident risks in US commercial nuclear power plants, NRC
18. WASH-740 (1957) Theoretical possibilities and consequences of major accidents in large nuclear power plants (The Brookhaven Report), AEC
19. German Risk Study (1979) Deutsche Risikostudie Kernkraftwerke, GRS, FRG
20. Brisbois J, Lanore JM, Villemeur A et al (1990) Les etudes probabilistes de surete des centrales nucleaires francaises de 900 et 1300 MWe
21. Severe accident risks: an assessment for five US nuclear power plants (1989) NUREG/CR-1150, NRC
22. Swaminathan S, Smidts C (1999) The mathematical formulation for the event sequence diagram framework. Rel Eng Syst Saf 65:103–118
23. Papazoglou IA (1998) Mathematical foundations of event trees. Rel Eng Syst Saf 61:169–183
24. Čepin M, Mavko B (2002) A dynamic fault tree. Rel Eng Syst Saf 75(1):83–91
25. Vrbanić I, Kaštelan M (1997) Optimization of NPP Krško PSA model structure by the employment of house events. Nuclear Energy in Central Europe, Proceedings, pp 414–421
26. IEEE Standard 500 (1984) IEEE Guide to the collection and presentation of electrical, electronic, sensing component, and mechanical equipment reliability data for nuclear-power generating stations, appendix D. Reliability Data for Nuclear-Power Generating Stations, IEEE
27. T-book: reliability data of components in nordic nuclear power plants (2000) Villingby, Sweden TUD Office and Pörn Consulting
28. T-Book (1992), ATV
29. IAEA-TECDOC-478 (1988) Component reliability data for use in probabilistic safety assessment. IAEA

30. Jordan Cizelj R, Mavko B, Kljenak I (2001) Component reliability assessment using quantitative and qualitative data. Rel Eng Syst Saf 71:81–95
31. Atwood CL, La Chance JL, Martz HF et al (2003) Handbook of parameter estimation for probabilistic risk assessment (NUREG/CR-6823). NRC
32. Measures of risk importance and their applications (1983) NUREG/CR-3385, NRC
33. Borgonovo E, Apostolakis GE (2001) A new importance measure for risk-informed decision making. Rel Eng Syst Saf 72:193–212
34. Čepin M (2010) Applications of the fault tree analysis for vulnerability studies (Chapter 8). In: Lesage A, Tondreau J (eds) Nuclear fuels: manufacturing processes, forms, and safety. Nova, New York
35. Čepin M, Cizelj L, Leskovar M, Mavko B (2006) Vulnerability analysis of a nuclear power plant considering detonations of explosive devices. J Nucl Sci Tech 43(10):1258–1269
36. Muthukumar CT, Guarro SB, Apostolakis G (1994) Dependability of embedded software systems, reliability and safety assessment of dynamic process systems. In: Aldemir T, Siu NS, Mosleh A, Cacciabue PC, Goktepe BG (eds) NATO ASI series F. Springer Verlag, Heidelberg, pp 59–77
37. Čepin M, Mavko B (1999) Fault tree developed by an object-based method improves requirements specification for safety-related systems. Rel Eng Syst Saf 63:111–125
38. Garrett J, Guarro SB, Apostolakis GE (1995) The dynamic flowgraph methodology for assessing the dependability of embedded software systems. IEEE Trans Syst Man Cybern 25(5):824–840
39. Vaurio JK (1995) Optimization of test and maintenance intervals based on risk and cost. Rel Eng Syst Saf 49:23–36
40. Čepin M, Mavko B (1997) Probabilistic safety assessment improves surveillance requirements in technical specifications. Rel Eng Syst Saf 56:69–77
41. Martorell S, Carlos S, Sanchez A, Serradell V (2000) Constrained optimization of test intervals using a steady-state genetic algorithm. Rel Eng Syst Saf 67:215–232
42. Čepin M (2002) Optimization of safety equipment outages improves safety. Rel Eng Syst Saf 77:71–80
43. Yang JE, Sung TY, Yin Y (2000) Optimization of the surveillance test interval of the safety systems at the plant level. Nucl Tech 132:352–365
44. Harunuzzaman M, Aldemir T (1996) Optimization of standby safety system maintenance schedules in nuclear power plants. Nucl Tech 113:354–367
45. Čepin M, Gomez Cobo A, Martorell S et al (1999) Methods for testing and maintenance of safety related equipment: examples from an IAEA research project. In: Proceedings of ESREL99: safety and reliability, pp 247–251
46. IAEA-TECDOC-669 (1992) Case study on the use of PSA methods: assessment of technical specifications for the reactor protection system instrumentation. IAEA, Vienna
47. Čepin M, Martorell S (2002) Evaluation of allowed outage time considering a set of plant configurations. Rel Eng Syst Saf 78:259–266
48. Siu N (1994) Risk assessment for dynamic systems: an overview. Rel Eng Syst Saf 43:43–73
49. Ren Y, Dugan JB (1998) Optimal design of reliable systems using static and dynamic fault trees. IEEE Trans Rel 234–244
50. Dugan JB (1991) Automated analysis of phased-mission reliability. IEEE Trans Rel 40(1):45–52
51. Dugan JB, Lyu MR (1994) System reliability analysis of an N-version programming application. IEEE Trans Rel 43(4):513–519
52. Burdick GR, Fussel JB, Rasmuson DM, Wilson JR (1977) Phased mission analysis: a review of new developments and an application. IEEE Trans Rel R 26(1):43–49
53. Modarres M, Cheon SW (1999) Function-centered modeling of engineering systems using the goal-success tree technique and functional primitives. Rel Eng Syst Saf 64:181–200
54. Hu YS, Modarres M (1999) Evaluating system behavior through dynamic master logic diagram modeling. Rel Eng Syst Saf 64:241–269

55. Matsuoka T, Kobayashi M (1988) GO-FLOW: a new reliability analysis methodology. Nucl Sci Eng 98:64–78
56. Farmer F (1967) Reactor safety and siting: a proposed risk criterion. Nucl Saf 8:539–548
57. Apostolakis GE (2004) How useful is quantitative risk assessment? Risk Anal 24:515–520
58. Berg HP, Gortz R, Schimetschka E (2003) Quantitative probabilistic safety criteria for licensing and operation of nuclear plants. BFS-SK-03/03, BFS
59. Čepin M (2007) The risk criteria for assessment of temporary changes in a nuclear power plant. Risk Anal 27(4):991–998
60. Caruso MA, Cheok MC, Cunningham MA et al (1999) An approach for using risk assessment in risk-informed decisions on plant-specific changes to the licensing basis. Rel Eng Syst Saf 63:231–242
61. Use of probabilistic risk assessment methods in nuclear activities: final policy statement (1995) Federal Register, NRC
62. Individual plant examination for severe accident vulnerabilities-10CFR 50.54(f) (1988) Generic Letter, GL 88-20, NRC
63. Criteria for the performance of probabilistic safety assessment applications (2002) GS-1.14, CSN
64. Safety assessment principles for nuclear plants (1992) Health & Safety Executive, London
65. RG 1.174 (2002) An approach for using probabilistic risk assessment in risk-informed decisions on plant-specific changes to the licensing basis, NRC
66. RG 1.177 (1998) An approach for plant-specific, risk-informed decision making: technical specifications, NRC
67. RG 1.200 (2007) An approach for determining the technical adequacy of probabilistic risk assessment results for risk-informed activities, NRC
68. RG 1.201 (2006) Guidelines for categorizing structures, systems, and components in nuclear power plants according to their safety significance, NRC
69. Probabilistic safety assessment (PSA) for nuclear power plants, regulatory standard (2005) S-294, Canadian Nuclear Safety Commission
70. Probabilistic safety analysis in safety management of nuclear power plants (2003) YVL-2.8, STUK
71. Holmberg J, Puikkinen U, Rosquist T, Simola K (2001) Decision criteria in PSA applications. NKS-44
72. Samanta P, Kim IS, Mankamo T, Vesely WE (1995) Handbook of methods for risk-based analyses of technical specifications (NUREG/CR-6141). NRC
73. TR-105396 (1995) PSA applications guide. Electric Power Research Institute
74. Martorell S, Carlos S, Villanueva JF, Sánchez AI et al (2006) Use of multiple objective evolutionary algorithms in optimizing surveillance requirements. Rel Eng Syst Saf 91(9):1027–1038
75. Keller W, Modarres M (2005) A Historical overview of probabilistic risk assessment development and its use in the nuclear power industry: a tribute to the late Professor Norman Carl Rasmussen. Rel Eng Syst Saf 89(3):271–285
76. NUREG/CR-1278 (1983) Handbook for human reliability analysis with emphasis on nuclear power plants application. NRC
77. Čepin M (2008) DEPEND-HRA: a method for consideration of dependency in human reliability analysis. Rel Eng Syst Saf 93(10):1452–1460
78. Čepin M (2007) Importance of human contribution within the human reliability analysis (IJS-HRA). J Loss Prev Proc Ind 21(3):268–276
79. Prošek A, Čepin M (2008) Success criteria time windows of operator actions using RELAP5/MOD33 within human reliability analysis. J Loss Prev Proc Ind 21(3):260–267
80. Volkanovski A, Čepin M, Mavko B (2009) Application of the fault tree analysis for assessment of power system reliability. Rel Eng Syst Saf 94(6):1116–1127

Chapter 6
Event Tree Analysis

When you have eliminated the impossible, whatever remains, however improbable, must be the truth

Arthur Conan Doyle

6.1 Introduction

Event tree analysis is the technique used to define potential accident sequences associated with a particular initiating event or set of initiating events [1–4]. The event tree model describes the logical connection between the potential successes and failures of defined safety systems or safety functions as they respond to the initiating event and the sequence of events [5–9].

6.2 Development Procedure

The sequences in the analyzed system show the success and the failure of the safety systems and actions available. The sequences include initiating event and the failures or successes of the safety functions. The event tree is a diagram that shows the initiating event and the failures or successes of the safety functions. The event tree is a standard tool in probabilistic risk assessment and it links the responses of safety systems following the initiating event. Figure 6.1 shows the general tasks included in the process of development of event trees:

- Plant familiarization
- Definition of safety functions and event tree headings
- Determination of system success criteria
- Identification of initiating events
- Definition of accident consequences
- Determination of plant damage state
- Event tree evaluation

Figure 6.2 shows an example of the event tree linked together with the fault tree. Abbreviation IE in Fig. 6.2 means initiating event. Abbreviation SS1 means safety system 1. Abbreviation OK means success state. Abbreviation CD means

M. Čepin, *Assessment of Power System Reliability*,
DOI: 10.1007/978-0-85729-688-7_6, © Springer-Verlag London Limited 2011

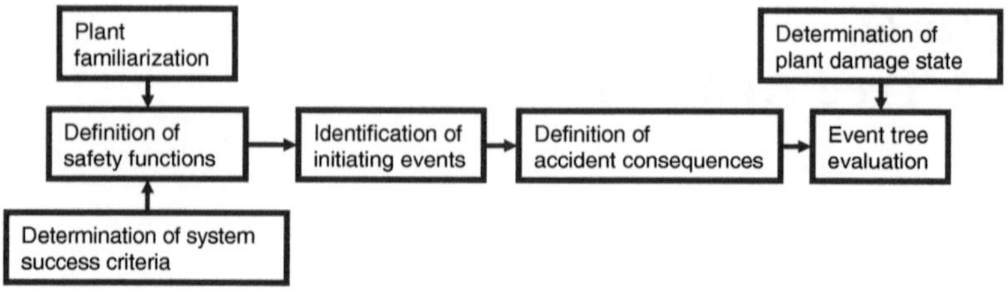

Fig. 6.1 Event tree development process

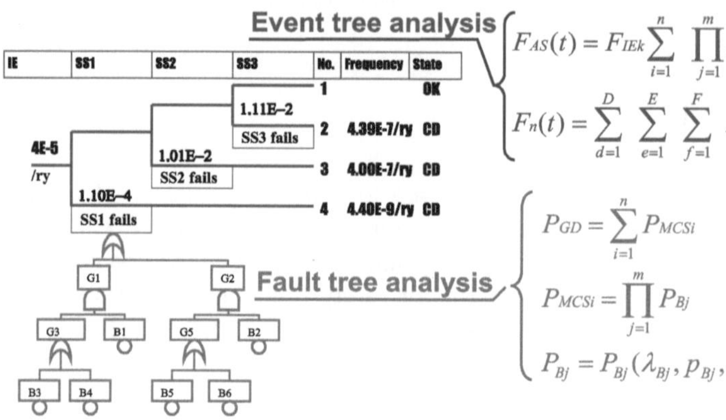

$$F_{AS}(t) = F_{IEk} \sum_{i=1}^{n} \prod_{j=1}^{m} P_{Bj}(t)$$

$$F_{n}(t) = \sum_{d=1}^{D} \sum_{e=1}^{E} \sum_{f=1}^{F} F_{ASdef}(t)$$

$$P_{GD} = \sum_{i=1}^{n} P_{MCSi}$$

$$P_{MCSi} = \prod_{j=1}^{m} P_{Bj}$$

$$P_{Bj} = P_{Bj}(\lambda_{Bj}, p_{Bj}, t_{mBj})$$

Fig. 6.2 Example of an event tree linked together with the fault tree

plant damage state: core damage (core damage of a reactor in a nuclear power plant, if the facility under investigation is nuclear power plant). Description of the example event tree from the figure continues at describing the phases of the event tree analysis [12–17].

6.3 Plant Familiarization

The purpose of the first task is to provide information necessary for the identification of initiating events, the identification of the success criteria for systems [16], which must directly perform the required safety functions, and the identification of the dependences between the frontline system and the support systems, which they require for proper functioning. The sources of information include the system description documents, design documents, drawings and procedures connected with the operability and testing, and maintenance of the systems.

6.4 Definition of Safety Functions and Event Tree Headings

The safety functions are the functions that must be performed to control the processes in the plant. For example, the safety functions in a nuclear power plant are defined by a group of actions that prevent core melting, prevent containment failure, or minimize radionuclide releases. Examples of safety functions that may appear as the headings in the event tree analysis are: high-pressure safety system injection, or auxiliary feedwater actuation, or emergency diesel generator start and operation, or operator restores the offsite power. For example, the safety functions in a chemical facility are the systems that prevent overpressures, excessive concentrations, and chemical releases to the environment. The event tree headings include safety systems and the success criteria. Figure 6.2 shows the event tree with three safety functions each represented by its respective safety system: SS1 or safety system 1, SS2 or safety system 2, and SS3 or safety system 3.

6.5 System Success Criteria

Success criterion is a term that defines the conditions for the system to operate successfully. For example, if the safety system includes three redundant portions of the system each consisting of a full power pump and several valves, it is necessary to define how many portions of the system are needed for the system to operate. If only one portion of the system is enough for the success of the system, the failure criteria, which suits the success criteria of the system, is failure of all three portions of the system.

One may say that the system models for the event tree headings require predefined failure criteria based on the success criteria defined for each event tree heading. The behavior of the safety system can be correlated with the behavior of the preceding safety system or not.

6.6 Identification of Initiating Events

The initiating event is an event that may jeopardize the safety of the plant if the safety systems do not prevent the undesirable consequences. The selection of initiating events consists of two steps:

- Definition of possible events
- Grouping of identified initial events by the safety function to be performed or combinations of systems responses

Examples of the initiating events include:

- Loss of offsite power (which means that the loss of outside sources of electrical energy happens)

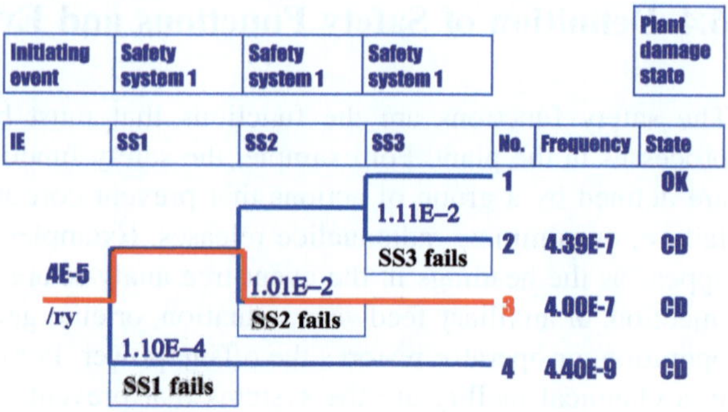

Fig. 6.3 Accident sequence of an event tree

- Large loss of a coolant accident (which means a break of a large pipe)
- Small loss of a coolant accident (which means a break of a small pipe).

6.7 Definition of Accident Consequences

Accident sequences represent the successes or failures of each particular safety system connected together with initiating event into a sequence of events. The headings of the event tree include the safety system functioning, which can be either failure or success. If the safety system succeeds, the accident sequence goes upward and if the safety system fails, the accident sequence goes downward.

Figure 6.3 shows one accident sequence in red color. This accident sequence is one out of four accident sequences of the event tree. The initiating event (IE) occurs. Then, the safety system 1 (SS1) succeeds, which means that the upward way of sequence is selected. Safety system 2 (SS2) fails, which means that the downward way of the sequence is selected. The safety system 3 is not represented with an option for success or failure in this particular accident sequence, because it does not matter if safety system 3 succeeds of fails, in both options, the plant damage state number 3 follows as the end state of this accident sequence. The selected accident sequence includes initiating event (IE), success of safety system 1 (SS1), and a failure of safety system 2 (SS2).

6.8 Determination of Plant Damage State

The final outcome of the accident sequence ends as the plant damage state. If the appropriate number of safety systems succeeded, the plant damage state is a safe state or sometimes named as OK state. If the safety systems failed, the plant damage state is a defined damage state. More damage states can be defined to distinguish in terms of nature of damage state or only one damage state is selected. Figures 6.2 and 6.3 show four accident sequences, each ends in its representative plant damage state. One accident sequence ends with success plant damage state, which is marked with OK. The other three accident sequences end with plant damage state core damage (CD), which means excessive temperature in the core of the reactor of nuclear power plant, for example.

6.9 Event Tree Evaluation

The event tree evaluation can be qualitative or quantitative or both. The evaluation is similar to the fault tree evaluation (see Chapter Fault Tree Analysis). Several variants of the evaluation exist, because the models can become very large and different approximations can be used. The event tree is evaluated in a way that each accident sequence is evaluated separately. When all accident sequences are evaluated, the evaluation of the event tree is finished, although the results of accident sequences can be further combined in groups of similar accident sequences and the common results can be of interest.

The most exact version of evaluation of the event tree is performed in a way that the de Morgan rules of Boolean algebra are applied in each particular fault tree that is evaluated as a part of the event tree evaluation and in joining the event tree branches together as a part of the accident sequence evaluation. Application of de Morgan rules causes that the negated events are considered in the analysis.

Simplifications of the event tree evaluation can include:

- Skipping of the de Morgan rules and thus the negated part of sequences within the event tree sequences.
- Simplifications in quantitative evaluations of the event trees, which can consider rare event approximation similarly as it is applied at the fault tree analysis or which can neglect the success failure probabilities of safety systems.

Qualitative evaluation of the event tree is described for the simple example shown in Fig. 6.4. Qualitative evaluation goes sequence by sequence. The qualitative evaluation of sequence 4 in Fig. 6.4 ends with state named CD. This accident sequence includes occurrence of initiating event (IE) and failure of safety system 1 (SS1). The qualitative evaluation is performed in a way that the fault tree representing SS1 failure is combined with initiating event.

$$\text{Accident sequence 4: AS4} = \text{IE} \times \text{SS1}$$

If the initiating event is not developed further in a fault tree, it is only one event and it is independent from any events from the evaluation of the fault tree SS1. In such case, the accident sequence is evaluated by the fault tree evaluation of the SS1 and the results are multiplied with initiating event IE.

The fault tree evaluation of SS1 gives the resulted expression, from which qualitative results, i. e., minimal cut sets are obtained:

$$\text{SS1} = B1 \times B3 + B1 \times B4 + B2 \times B5 + B2 \times B6 \tag{6.1}$$

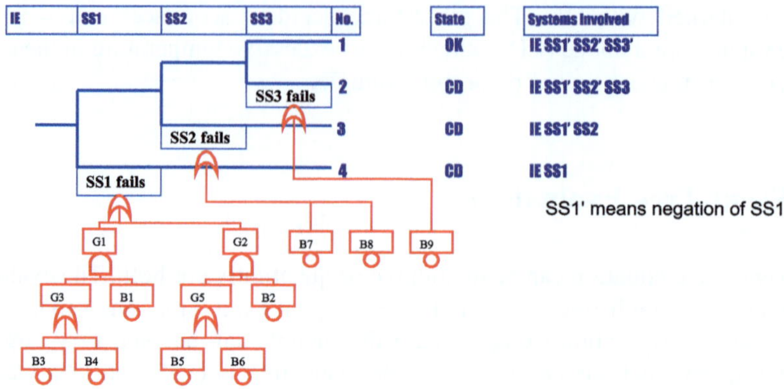

Fig. 6.4 Qualitative evaluation of an event tree

Qualitative result of accident sequence 4 is derived from SS1 fault tree and IE event:

$$\begin{aligned} \mathrm{AS4} &= \mathrm{IE} \times \mathrm{SS1} \\ &= \mathrm{IE} \times B1 \times B3 + \mathrm{IE} \times B1 \times B4 + \mathrm{IE} \times B2 \times B5 + \mathrm{IE} \times B2 \times B6 \end{aligned} \tag{6.2}$$

Qualitative result of accident sequence 3 is derived from considering the negation of failure of SS1, because SS1 succeeds in this sequence, considering fault tree of safety system SS2, because safety system 2 fails in this sequence, and considering initiating event (IE):

$$\begin{aligned} \mathrm{AS3} &= \mathrm{IE} \times \mathrm{SS1}' \times \mathrm{SS2} \\ &= \mathrm{IE} \times (B1 \times B3 + B1 \times B4 + B2 \times B5 + B2 \times B6)' \times (B7 + B8) \end{aligned} \tag{6.3}$$

The exact evaluation needs application of de Morgan rules and it gives complex expression for relatively simple problems and it consequently causes that in the minimal cut sets several or many negated events exists. This reduces significantly the clear picture about the qualitative results.

Therefore, the simplification can be performed, which neglects the successful safety system 1 occurrence and the simplified expression gives:

$$\mathrm{AS3} = \mathrm{IE} \times \mathrm{SS2} = \mathrm{IE} \times (B7 + B8) = \mathrm{IE} \times B7 + \mathrm{IE} \times B8 \tag{6.4}$$

Similarly, the expressions for accident sequence 2 are the following, exact:

$$\begin{aligned} \mathrm{AS2} &= \mathrm{IE} \times \mathrm{SS1}' \times \mathrm{SS2}' \times \mathrm{SS3} \\ &= \mathrm{IE} \times (B1 \times B3 + B1 \times B4 + B2 \times B5 + B2 \times B6)' \times (B7 + B8)' \times B9 \end{aligned} \tag{6.5}$$

and simplified:

$$\mathrm{AS2} = \mathrm{IE} \times \mathrm{SS3} = \mathrm{IE} \times B9 \tag{6.6}$$

The sequence 1 is not an accident sequence as it ends in a successful state, it is only a sequence, so the evaluation may not be so important, although it can be done:

$$\begin{aligned} \mathrm{AS1} &= \mathrm{IE} \times \mathrm{SS1}' \times \mathrm{SS2}' \times \mathrm{SS3}' \\ &= \mathrm{IE} \times (B1 \times B3 + B1 \times B4 + B2 \times B5 + B2 \times B6)' \times (B7 + B8)' \times B9' \end{aligned} \tag{6.7}$$

When all accident sequences have been qualitatively evaluated, some grouping of similar accident sequences through one event tree or through several event trees can be performed. For example, from Fig. 6.4, the state CD is of interest, which is the outcome of accident sequences 2, 3, and 4:

$$\begin{aligned} \mathrm{CD} &= \mathrm{AS2} + \mathrm{AS3} + \mathrm{AS4} \\ &= \mathrm{IE} \times B9 + \mathrm{IE} \times B7 + \mathrm{IE} \times B8 + \mathrm{IE} \times B1 \times B3 + \mathrm{IE} \times B1 \times B4 + \mathrm{IE} \times B2 \\ &\quad \times B5 + \mathrm{IE} \times B2 \times B6 \end{aligned} \tag{6.8}$$

The qualitative evaluation of the event tree gives the minimal cut sets for the accident sequences or the minimal cut sets for the plant damage states.

$$\mathrm{AS} = \mathrm{IE} \times \sum_{i=1}^{I} \prod_{j=1}^{J} Bj \tag{6.9}$$

where IE is the initiating event, B_j is the basic event j, J is the number of basic events in a particular minimal cut set, and I number of minimal cut sets.

Quantitative evaluation of the event tree is performed similarly as the quantitative evaluation of the fault tree. The events from the qualitative expression have to be equipped with their failure probabilities. As the initiating events can be quantified in terms of probability or frequency, the results of the event tree evaluation can be expressed as frequency f_{IE} or probability p_{IE} depending on the input from initiating events. The results of quantitative evaluation of the event tree are shown assuming the initiating event frequency:

- The frequencies of accident sequences
- The frequencies of the groups of accident sequences
- The overall frequency of the postulated accident, which is calculated as the sum of all corresponding accident sequences
- Importance factors, such as Fussel–Vesely importance, risk achievement worth, risk reduction worth, Birnbaum importance, and criticality importance

$$\begin{aligned} f_{\mathrm{AS}} = f_{\mathrm{IE}} \Bigg(& \sum_{i=1}^{n} P_{\mathrm{MCS}_i} - \sum_{i<j} P_{\mathrm{MCS}_i \cap \mathrm{MCS}_j} + \sum_{i<j<k} P_{\mathrm{MCS}_i \cap \mathrm{MCS}_j \cap \mathrm{MCS}_k} - \cdots \\ & + (-1)^{m-1} P \bigcap_{i=1}^{m} \mathrm{MCS}_i \Bigg) \end{aligned} \tag{6.10}$$

where f_{AS} is the accident sequence frequency, f_{IE} is the initiating event frequency, $P_{\mathrm{mcs}i}$ is the probability of occurrence of minimal cut set i (MCS_i), n is the number of minimal cut sets, and m is the number of basic events in the largest minimal cut set.

$$P_{\mathrm{MCS}_i} = P_{B1} \times P_{B2}|P_{B1} \times P_{B3}|P_{B1} \cap P_{B2}, \ldots, P_{Bm}|P_{B1} \cap P_{B2} \cap, \ldots, \cap P_{Bm-1} \tag{6.11}$$

Parameters P_{B1}, P_{B2}, ..., P_{Bn} represent the failure probabilities of basic events: $B1$, $B2$, ..., Bn, respectively. The probabilities of basic events are calculated for each basic event using the selected probabilistic model.

Under the assumption that the basic events are mutually independent, the following equation stands.

$$P_{\mathrm{MCS}_i} = \prod_{j=1}^{m} P_{Bj} \tag{6.12}$$

Failure probability of each basic event is a function of respective parameters from the selected probabilistic model. The general equation for all probabilistic models is represented by the following equation, where the failure probability of the respective basic event is a function of only certain parameters, which are connected with the respective probabilistic model.

$$P_{Bj} = P_{Bj}\left(\lambda_{Bj}, p_{Bj}, T_{iBj}, T_{tBj}, T_{mBj}, T_{rBj}, \eta_{Bj}\right) \tag{6.13}$$

where λ_{Bj} is the failure rate of equipment modeled in basic event Bj, p_{Bj} is the probability per demand of equipment modeled in basic event Bj, T_{iBj} is the test interval of equipment modeled in basic event Bj, T_{tBj} is the test duration of equipment modeled in basic event Bj, T_{mBj} is the mission time of equipment modeled in basic event Bj, T_{rBj} is the repair time of equipment modeled in basic event Bj, and μ_{Bj} is the repair rate of equipment modeled in basic event Bj.

Sometimes, a first-order approximation is used.

$$f_{\mathrm{AS}} = f_{\mathrm{IE}}\left(\sum_{i=1}^{n} P_{\mathrm{MCS}_i}\right) \tag{6.14}$$

The quantitative results of accident sequences can be grouped over the selected groups of accident sequences or over all accident sequences.

$$f_{\mathrm{PLANT\_DAMAGE\_STATE}} = \sum_{z=1}^{Z} F_{\mathrm{AS}} = f_{\mathrm{IE}} \sum_{z=1}^{Z} \left(\sum_{i=1}^{n} P_{\mathrm{MCS}_i}\right) \tag{6.15}$$

where Z is the number of accident sequences that form the plant damage state.

Several event trees can be considered for several initiating events for one plant and the grouping of the frequencies can also be done over several event trees

$$f_{\mathrm{PLANT\_DAMAGE\_STATE\_ETs}} = \sum_{v=1}^{V} \sum_{z=1}^{Z} F_{\mathrm{AS}} = \sum_{v=1}^{V} f_{\mathrm{IEv}} \sum_{z=1}^{Z} \left(\sum_{i=1}^{n} P_{\mathrm{MCS}_i}\right) \tag{6.16}$$

where $f_{\mathrm{PLANT\_DAMAGE\_STATE\_ETs}}$ is the frequency of plant damage state over several event trees, V is the number of the event trees and the number of initiating events, f_{IEv} is the frequency of initiating event v.

If the initiating event probability is used for the quantification purposes instead of initiating event frequency, the equations are similar to equations above.

The equations for calculation of importance factors are similar to their respective equations from the fault tree analysis. The difference lays in the initiating event frequencies, which are part of the minimal cut sets in the event tree evaluation and are not part of the minimal cut sets in the fault tree evaluation.

Fussel–Vesely importance shows the contribution of the event to the accident sequence frequency on the level of accident sequences. It provides a numerical significance of all the events as parts of the accident sequence and allows them to be prioritized. It is calculated according to the following equation.

$$\mathrm{FV}_k = 1 - \frac{f_{\mathrm{AS}}(P_k = 0)}{f_{\mathrm{AS}}} = 1 - \frac{1}{\mathrm{RRW}_k} \tag{6.17}$$

where FV_k is the Fussel–Vesely importance for component modeled in basic event k, f_{AS} is the accident sequence frequency, $f_{\mathrm{AS}}(P_k = 0)$ is the accident sequence frequency when failure probability of component modeled in basic event k is set to 0, and RRW_k is the risk reduction worth for component k.

The risk achievement worth for a basic event shows the increase in the accident sequence frequency that would be obtained if the lower-level event, e.g., the failure of a component modeled in the basic event, would occur.

$$\mathrm{RAW}_k = \frac{f_{\mathrm{AS}}(P_k = 1)}{f_{\mathrm{AS}}} \tag{6.18}$$

where RAW_k is the risk achievement worth for component modeled in basic event k, $f_{\mathrm{AS}}(P_k = 1)$ is the accident sequence frequency when failure probability of component modeled in basic event k is set to 1, and f_{AS} is the accident sequence frequency.

The risk reduction worth for a basic event shows the decrease in the accident sequence frequency that would be obtained if the lower-level event, e.g., the failure of a component modeled in the basic event, did not occur. The large risk reduction worth implies that the risk of the respective basic event is worth to decrease in order that the overall risk is significantly decreased.

$$\mathrm{RRW}_k = \frac{f_{\mathrm{AS}}}{f_{\mathrm{AS}}(P_k = 0)} \tag{6.19}$$

where f_{AS} is the accident sequence frequency, $f_{\mathrm{AS}}(P_k = 0)$ is the accident sequence frequency when failure probability of component modeled in basic event k is set to 0, and RRW_k is the risk reduction worth for component modeled in basic event k.

The Birnbaum importance represents the rate of change in the accident sequence frequency as a result of the change in the probability of a given event.

The Birnbaum importance can be calculated by first calculating the accident sequence frequency with the probability of the given event set to 1 and then subtracting the accident sequence frequency with the probability of the given event set to 0.

$$B_k = f_{AS}(P_k = 1) - f_{AS}(P_k = 0) \tag{6.20}$$

where $f_{AS}(P_k = 1)$ is the accident sequence frequency when failure probability of component modeled in basic event k is set to 1 and $f_{AS}(P_k = 0)$ is the accident sequence frequency when failure probability of component modeled in basic event k is set to 0.

The criticality importance represents the rate of change in the accident sequence frequency as a result of the change in the probability of a given event relatively to the accident sequence frequency and failure probability of component modeled in basic event k.

$$C_k = \frac{f_{AS}(P_k = 1) - f_{AS}(P_k = 0)}{f_{AS}} P_k \tag{6.21}$$

where $f_{AS}(P_k = 1)$ is the accident sequence frequency when failure probability of component modeled in basic event k is set to 1, $f_{AS}(P_k = 0)$ is the accident sequence frequency when failure probability of component modeled in basic event k is set to 0, f_{AS} is the accident sequence frequency, and P_k is the failure probability of component k.

6.10 Linking of Event Trees With Fault Trees

Each heading of the event tree can be either quantified using detailed system models to determine the likelihood of system failure, which is mostly the case of safety systems, or assessed as one failure probability, which is mostly the case for the headings of the event tree connected with certain operator actions.

Two general methods or approaches exist for the event tree linking process with the fault tree analysis:

- Small event tree and large fault tree approach
- Large event tree and small fault tree approach

The small event tree and large fault tree approach is mostly used in nuclear industry in probabilistic safety assessment of nuclear power plants. The smaller number of safety systems is in the event tree headings and more detailed systems analyses are performed with the fault trees [8, 9]. Examples of the event trees are presented in references [13–15, 17].

References

1. Papazoglou IA (1998) Mathematical foundations of event trees. Rel Eng Syst Saf 61:169–183
2. Haasl D, Young J, Cramond WR (1985) Probabilistic risk assessment course documentation: system reliability and analysis techniques session. NUREG/CR-4350, vol 4. NRC, Washington
3. Knief RA (1992) Nuclear engineering: theory and technology of commercial nuclear power. Hemisphere, New York
4. Kumamoto H, Henley EJ (1996) Probabilistic risk assessment and management for engineers and scientists. IEEE, New York
5. ASME RA-S-2002 (2002) Standard for probabilistic risk assessment for nuclear power plant applications. Addendum (2005), ASME
6. RA-S-2008 (2008) Standard for level 1/large early release frequency probabilistic risk assessment for nuclear power plant applications. ASME
7. Villemeur A (1992) Reliability, availability, maintainability and safety assessment: methods and techniques. Wiley, New York
8. Probabilistic Risk Assessment Procedures Guide (1982) NUREG/CR-2300, NRC
9. Probabilistic Safety Analysis Procedures Guide (1985) NUREG/CR-2815, NRC
10. PRA NASA Guide (2002) Probabilistic risk assessment procedures guide for NASA managers and practitioners. NASA
11. Čepin M (2005) Analysis of truncation limit in probabilistic safety assessment. Rel Eng Syst Saf 87(3):395–403
12. Analysis of Core Damage Frequency (1990) NUREG/CR-4550, NRC
13. WASH-1400 (1975) Reactor safety study: an assessment of accident risks in US commercial nuclear power plants. NRC
14. German Risk Study (1979) Deutsche Risikostudie Kernkraftwerke. GRS, FRG
15. US NRC (1989) Severe accident risks: an assessment for five US nuclear power plants (NUREG/CR-1150). NRC
16. Prior RP, Chaboteaux JP, Wolvaardt FP et al (1994) Best estimate success criteria in the Krško IPE. In: International meeting on PSA/PRA and Severe Accidents, NSS, Ljubljana
17. Lungmen Units 1&2 (2006) Preliminary safety analysis report

Chapter 7
Binary Decision Diagram

It is a truth very certain that when it is not in our power to determine what is true, we ought to follow what is most probable

Rene Descartes

7.1 Introduction

A binary decision diagram (BDD) is a mean to represent, analyze, test, and implement Boolean functions [1–5]. It is a directed acyclic graph that consists of nodes and edges [6–9]. The popular description of using such diagrams to represent Boolean functions was provided by Akers in 1978 [2]. Although other methods may be used to complete such tasks, e.g., the Karnaugh map, binary decision diagrams offer some useful advantages [10–20]. Karnaugh maps and truth tables are suitable methods that may be used to describe functions consisting of a small number of variables [3]. However, the problem with such methods is that the size of these structures increases dramatically as the number of variables increase, i.e., 2^n rows in a truth table or squares in a Karnaugh map are required for a function of n variables. Although a binary decision diagram contains $2^n - 1$ nodes for n variables, it is possible to reduce the size of these structures by following certain algorithms. For instance, the order in which variables are evaluated within a binary decision diagram can significantly affect the size of the structure. A binary decision diagram is called an ordered binary decision diagram (OBDD) if on all of its paths the variables appear in the same order [7].

By implementing such ordering restrictions algorithms, a more efficient manipulation of the diagrams is possible [4]. This additional ordering also allows for a reduction technique to be applied to the ordered binary decision diagram to remove redundancies from the data structure and therefore produce a more compact representation of the Boolean expression. Such structures are known as reduced ordered binary decision diagrams (ROBDD). In addition, the reduced ordered binary decision diagrams provide a unique representation of a Boolean function [2].

This technique is able to produce a completely unique graphical representation of a Boolean function that in turn may be used to test for equivalence of a Boolean function. Advantages of reduced ordered binary decision diagrams include:

M. Čepin, *Assessment of Power System Reliability*,
DOI: 10.1007/978-0-85729-688-7_7,

- Compact representation of Boolean functions
- A canonical representation of Boolean functions
- Efficient manipulation of functions expressed by reduced ordered binary decision diagrams

In consistent with the current literature, the reduced ordered binary decision diagrams will simply be referred to as binary decision diagrams. Binary decision diagrams are a type of direct acyclic graph [2]. An ordered binary decision diagram (OBDD) was achieved by placing restrictions on the order in which Boolean variables in the diagram were evaluated [4]. A binary decision diagram consists of:

- A set of decision nodes, starting at the root node at the top of the decision diagram. Each decision node contains two outgoing branches, one is a high (1) branch and the other is a low (0) branch. These branches may be represented as *solid* and *dotted lines*, respectively.
- High and low branches are used to connect decision nodes with each other to create decision paths, i.e., based on the evaluation of the decision node, either the high or low branches are followed to the next decision node.
- The high and low branches of the final decision nodes are connected to either a high- (1) or low-terminal (0) node, which represents the output of the function.

7.2 Constructing a Binary Decision Diagram from a Simple Boolean Equation

Constructing a binary decision diagram from a simple Boolean equation represented by the function F is the following.

$$F = A' + BC' \tag{7.1}$$

To analyze the output of the function F, one must first consider the possible values that the Boolean variables (A, B, and C) may take. For example, if $A = 0$, then it is immediately clear that $F = 1$. However, if $A = 1$, then the values of B and C must also be considered. In this case, if $B = 0$, then it is clear that $F = 0$. Otherwise, if $B = 1$, then the value of C must be considered. In this case, it is clear that F will take on the value of C (i.e., given that $A = 0$ and $B = 1$, then if $C = 0$, then $F = 1$ or if $C = 1$, then $F = 0$).

Such an analysis is shown graphically in Fig. 7.1. The analysis begins at the top of the diagram with the decision node (represented by a *circle*) for the variable A. Based on the value of A (either 0 or 1), a particular branch is followed (either *a dashed* or *solid line*). When a terminal node (represented by *a square*) is reached, the output of the function F (either 0 or 1) is given based on the path followed through the diagram. As one can see, this method provides a clear graphical representation of the analysis described above. The resulting diagram is referred to

Fig. 7.1 Binary decision diagram for a simple Boolean equation

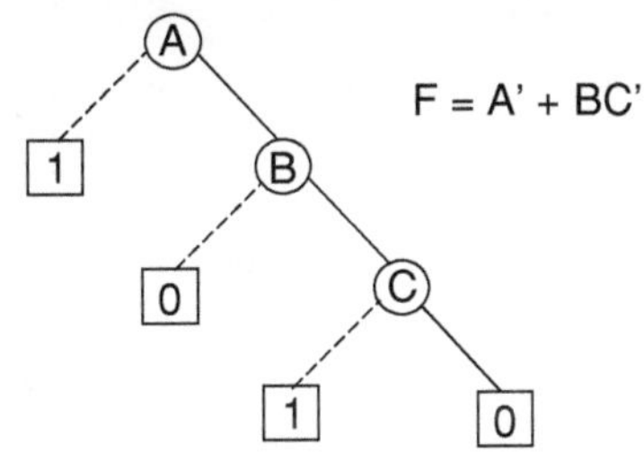

as a binary decision diagram [2]. Figure 7.1 demonstrates how a binary decision diagram may be constructed given a simple Boolean equation.

7.3 Constructing a Binary Decision Diagram from a Truth Table

Boolean functions may also be presented in other forms such as in a truth table. A truth table contains 2^n rows (n being the number of variables in the Boolean function) and therefore it is not a compact representation of a Boolean function. Similarly, constructing a binary decision diagram from a truth table leads to a large diagram containing 2^n paths from the starting node to the terminal nodes, resulting in $2^n - 1$ decision nodes. As one can imagine, large values of n lead to cumbersome representations of Boolean functions.

Figure 7.2 shows a binary decision tree, which is quite similar to a binary decision diagram. However, one point of difference is that a binary decision diagram contains nodes that may have more than one branch attached. Table 7.1 contains the truth table. Figure 7.2 shows the binary decision diagram that contains 16 paths (2^n with $n = 4$) representing the possible outputs of the function described by F.

$$F = A'BC'D' + A'BC'D + ABC'D' + ABC'D + ABCD' + AB'CD' \tag{7.2}$$

As one can imagine, with an increased number of variables (large n), the resultant binary decision diagram becomes very large in size. However, it is often the case with large binary decision diagrams (Fig. 7.2, for example) that there is redundant information that may be removed to produce a more compact diagram.

This may be achieved by following some simple reduction algorithms described in the following. Two simple steps may be used to obtain a reduced ordered binary decision diagram, known as (i) elimination and (ii) merging (Fig. 7.3).

Elimination: If the 0 and 1 branches of a node, x, are both connected to a lower node y, then the node x may be removed from the diagram (eliminated) and any incoming edges are connected to y.

Merging: If two nodes labeled by the same variable, x, have 1 and 0 branches that are connected to the same subsequent nodes, then the two nodes may be reduced to one node (merged).

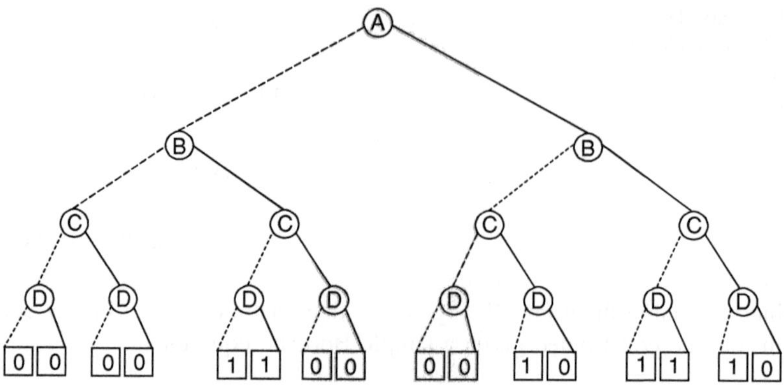

F = A'BC'D' + A'BC'D + ABC'D' + ABC'D + ABCD' + AB'CD'

Fig. 7.2 Binary decision diagram for a truth table

Table 7.1 Truth table

A	B	C	D	F
0	0	0	0	0
0	0	0	1	0
0	0	1	0	0
0	0	1	1	0
0	1	0	0	1
0	1	0	1	1
0	1	1	0	0
0	1	1	1	0
1	0	0	0	0
1	0	0	1	0
1	0	1	0	1
1	0	1	1	0
1	1	0	0	1
1	1	0	1	1
1	1	1	0	1
1	1	1	1	0

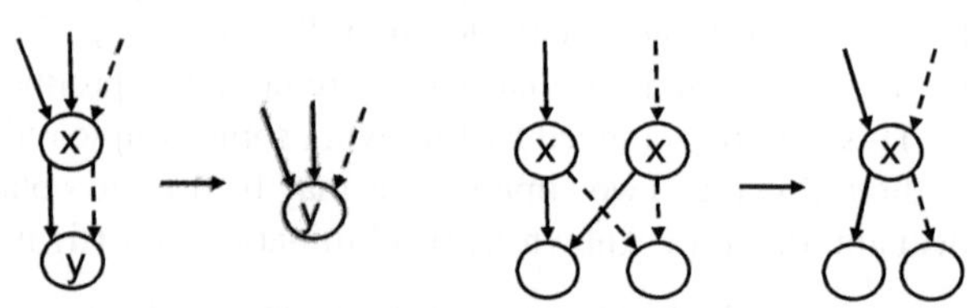

Fig. 7.3 Elimination and merging rules for a binary decision diagram

Both steps are repeated in turn until there are no redundancies left in the diagram. The reduced form of binary decision diagram is known as a reduced ordered binary decision diagram. The word *ordered* arises from the ordering of variables presented in the binary decision diagram and is described in more details further on.

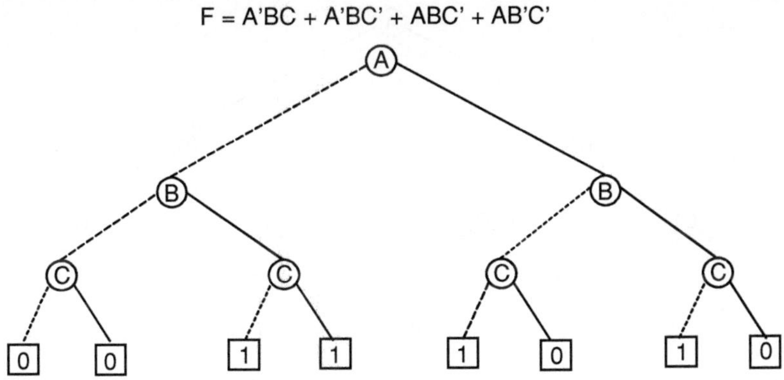

Fig. 7.4 Binary decision diagram example

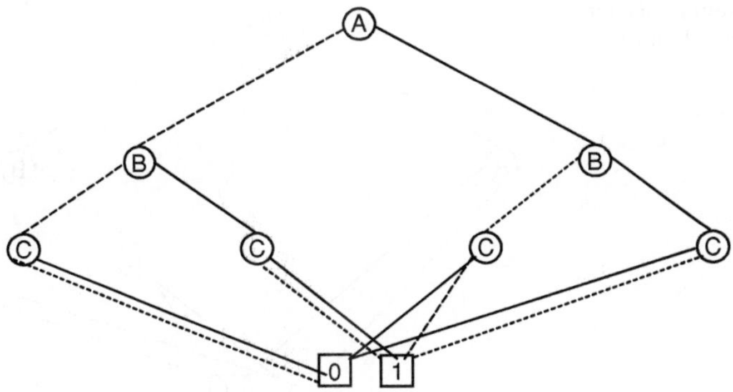

Fig. 7.5 Terminal nodes merged together

7.4 Reducing a Binary Decision Diagram to a More Compact Form

Figure 7.4 shows an example of binary decision diagram. Figure 7.5 shows the first step of reduction of the example binary decision diagram, where the common terminal nodes are merged together so there is only a single 0 terminal node and a single 1 terminal node.

In the next step, the nodes to be eliminated are identified and the branches from the previous node are re-connected appropriately. Figure 7.6 shows the marked and Fig. 7.7 shows other than eliminated nodes.

Figure 7.8 shows two nodes on the *right side* of the binary decision diagram marked with *red ellipse*, which share common destinations for their both 0 and 1 branches. As such, these nodes may be merged into a common node and the branches from the previous node are reconnected. In this example, merging these two nodes together has generated another node to be eliminated (Fig. 7.9).

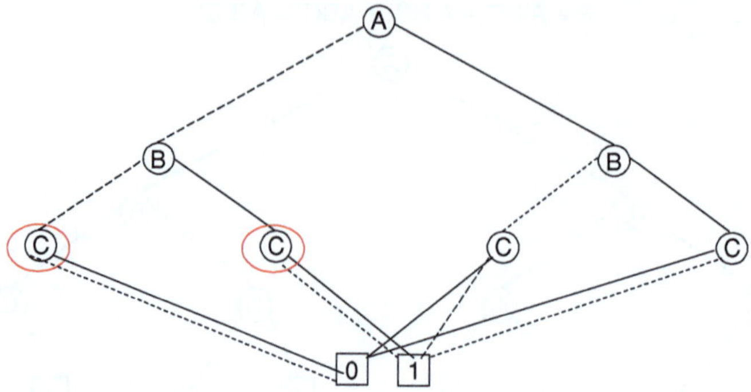

Fig. 7.6 Nodes to be eliminated are marked by *red circles*

Fig. 7.7 Resulting diagram after nodes are eliminated

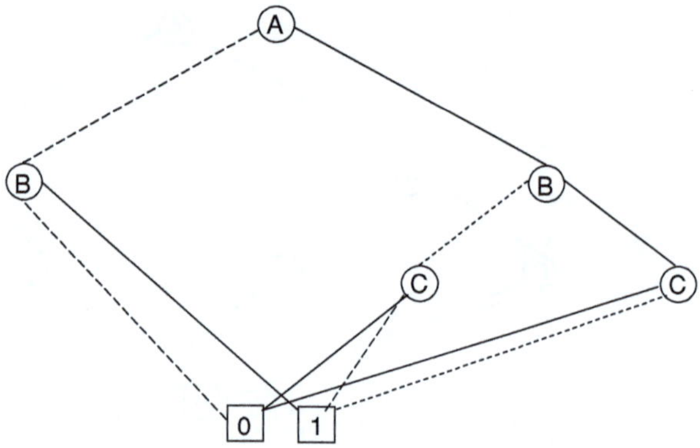

Fig. 7.8 Nodes to be merged are identified

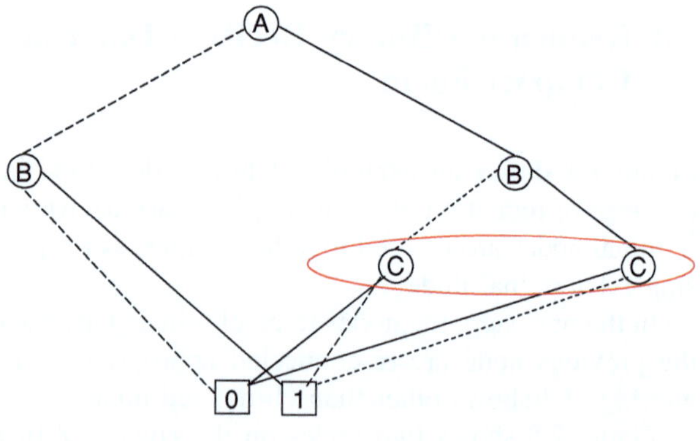

After the elimination of this node, there are no further eliminations or merges that may be performed and therefore the binary decision diagram has been reduced as shown on Fig. 7.10.

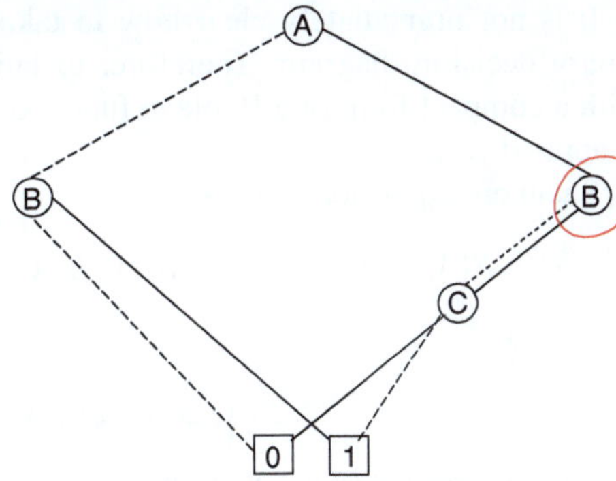

Fig. 7.9 Merging of the two nodes results in another node to be eliminated

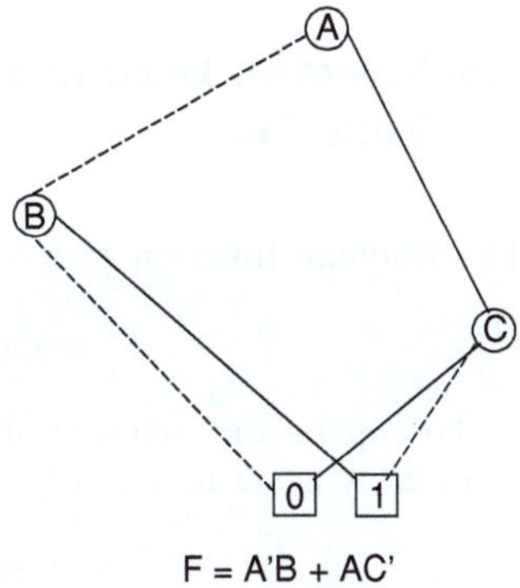

Fig. 7.10 Completely reduced binary decision diagram

Reading the reduced binary decision diagram provides a simplified Boolean expression that is equivalent to the original expression. This reduced binary decision diagram is considered canonical which means that it is a unique representation of the Boolean function which it describes [2, 4]. However, it should be noted that the reduced expression obtained by this method is not necessarily the most simplified expression, i.e., the implicants obtained are not necessarily prime implicants [2].

7.5 Obtaining a Binary Decision Diagram Using Shannon Decomposition

As it has already been shown, it is simple to construct a binary decision diagram from a truth table or a simple (small n) Boolean expression. However, Boolean functions are often provided in more compact forms that may not be immediately converted into a binary decision diagram, for example, function F:

$$F = B(A'C + C'E') + E'(A'B + B'D) \tag{7.3}$$

It is not immediately clear how to take such a function and represent it in a binary decision diagram. Therefore, to build a binary decision diagram starting with a compact form of a Boolean function, the Shannon decomposition technique is applied [2].

Shannon expansion method:

$$F(A, B, C, D\ldots) = AF(1, B, C, D\ldots) + A'F(0, B, C, D\ldots) \tag{7.4}$$

or

$$F(A, B, C, D\ldots) = AF_A + A'F_{A'} \tag{7.5}$$

That is, any Boolean function may be re-written in terms of two sub-functions.

7.6 Shannon Decomposition of a Five-Variable Boolean Function

The Boolean function F is considered.

$$F = C(A'E + B'E') + D'(A'B + B'C) \tag{7.6}$$

To expand the function about the variable A:
First, A is set to 1 to obtain:

$$F(A = 1) = F_A = B'CE' + B'CD' \tag{7.7}$$

Second, A is set to 0 to obtain:

$$F(A = 0) = F_{A'} = C(E + B'E') + D'(B + B'C) \tag{7.8}$$

With further simplification:

$$F_{A'} = C(E + B') + D'(B + C) \tag{7.9}$$

Because:

$$x + x'y = x + y \tag{7.10}$$

To produce a binary decision diagram from this method, one must work from the top-down and repeat the Shannon expansion formulae for each new function obtained [2].

7.7 Creating a Binary Decision Diagram Using Repeated Shannon Decomposition

Figures 7.11, 7.12, 7.13, 7.14 and 7.15 show Shannon expansion of function F by the variables A, B, C, D, and E, respectively.

Fig. 7.11 Shannon expansion of function F about variable A

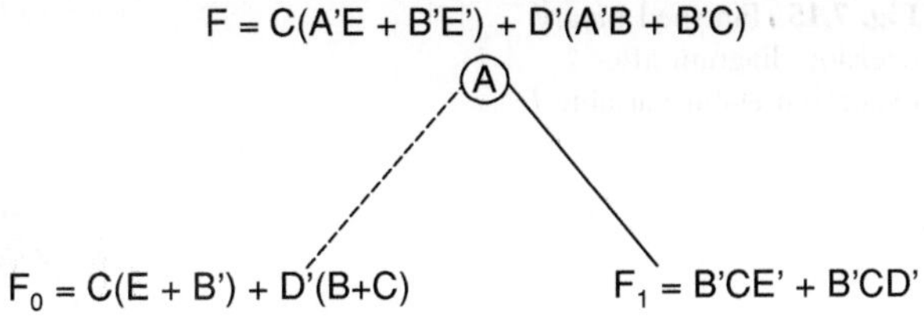

Fig. 7.12 Shannon expansion of F_0 and F_1 about variable B

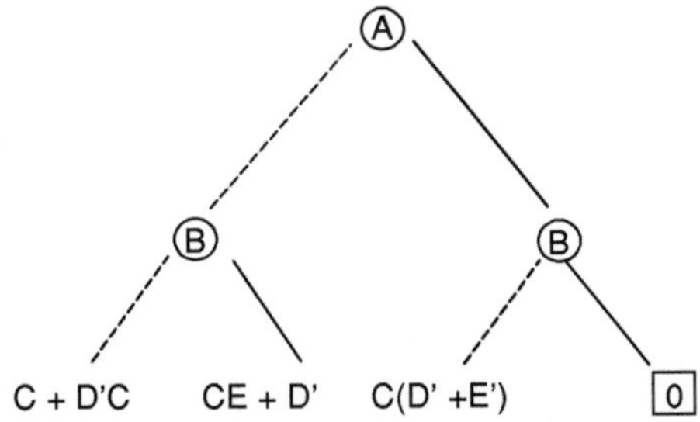

Fig. 7.13 Shannon expansion about variable C

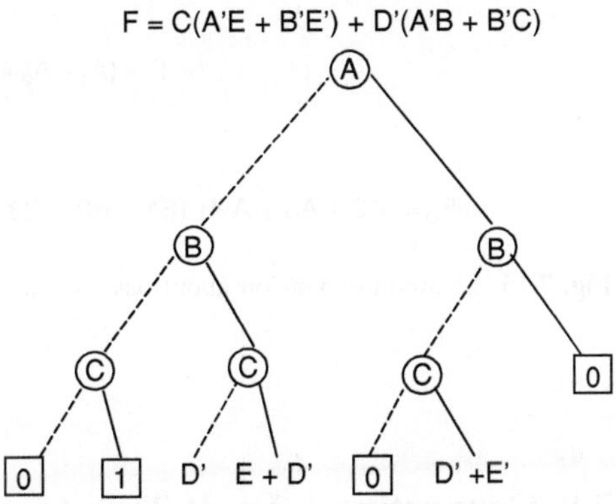

Fig. 7.14 Shannon expansion about variable D

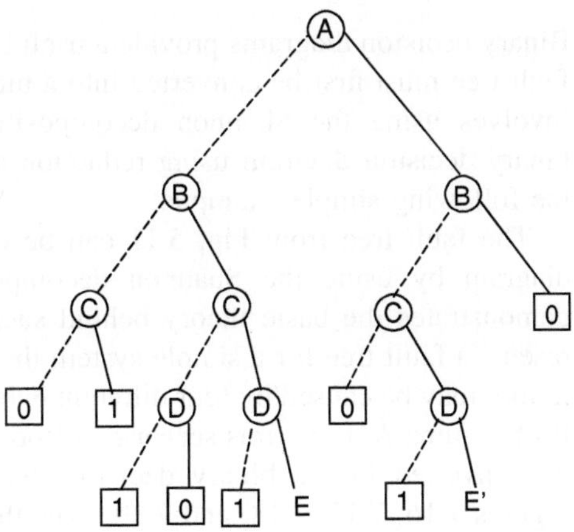

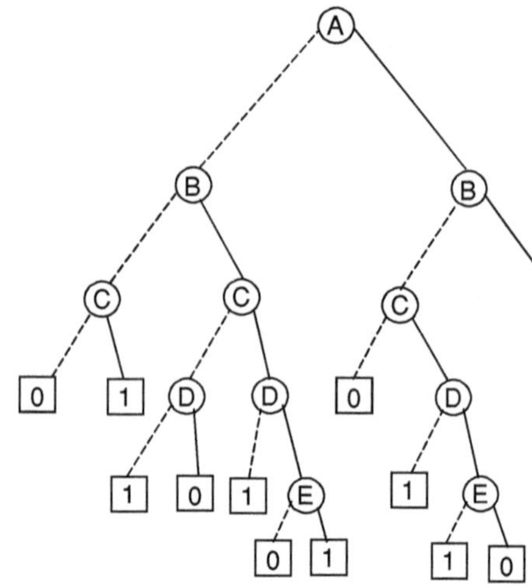

Fig. 7.15 Finished binary decision diagram after expansion about variable *E*

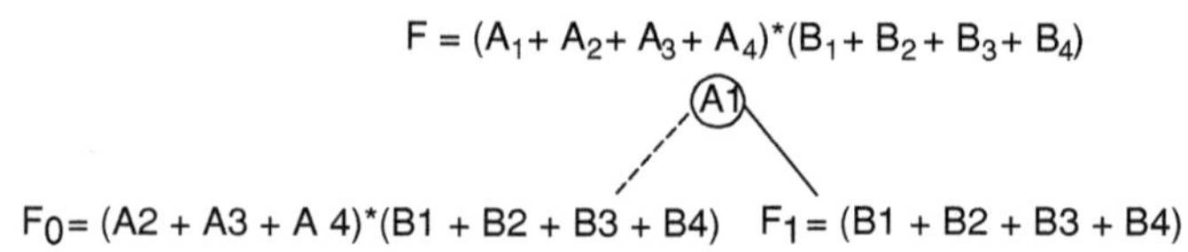

Fig. 7.16 Shannon expansion about basic event A1

7.8 Converting a Fault Tree to a Binary Decision Diagram

Binary decision diagrams provide a useful tool to perform fault tree analyses. The fault tree must first be converted into a binary decision diagram [6]. This process involves using the Shannon decomposition followed by simplification of the binary decision diagram using reduction algorithms [6]. This is demonstrated in the following simple example.

The fault tree from Fig. 5.11 can be easily converted into a binary decision diagram by using the Shannon decomposition method. This simple example demonstrates the basic theory behind such an implementation. Figure 5.11 represents a fault tree for a simple system that may fail if pumps *A* and *B* fail. Pump failure may be caused by four different events labeled A1–A4 for pump *A* and B1–B4 for pump *B*. From this scenario, a Boolean expression is derived that may then be converted into a binary decision diagram using Shannon's decomposition. Figures 7.16, 7.17, 7.18, and 7.19 show this process.

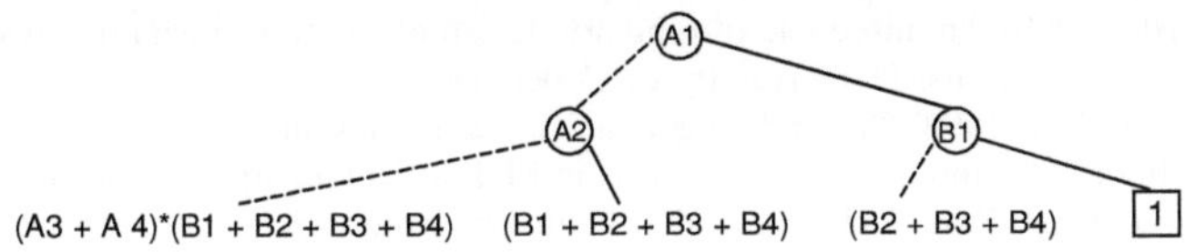

Fig. 7.17 Shannon expansion about basic event A2 and B1

Fig. 7.18 Shannon expansion about basic event A3 and B2

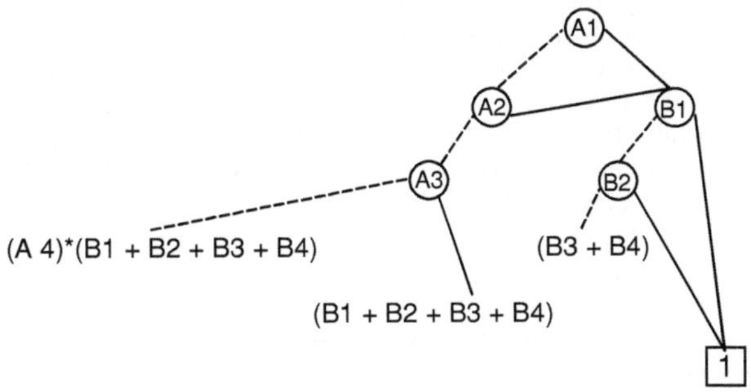

Fig. 7.19 Finished binary decision diagram after expansion

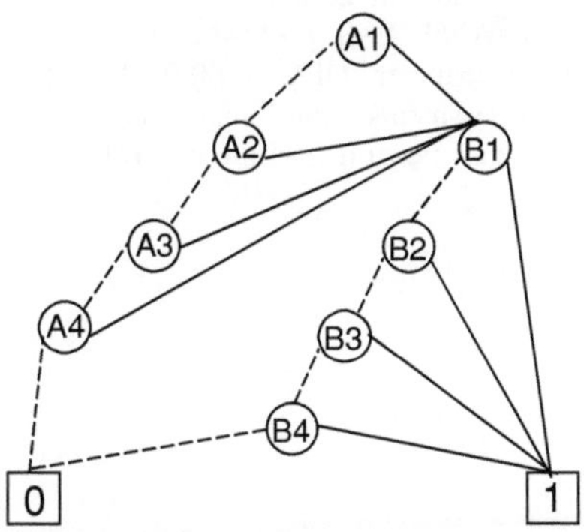

References

1. Lee CY (1959) Representation of switching functions by binary decision programs. Bell Syst Tech J 38:985–999
2. Akers SB (1978) Binary decision diagrams. IEEE Trans Comput c-27(6):509–516
3. Rushdi AM (1983) Symbolic reliability analysis with the aid of variable-entered Karnaugh maps. IEEE Trans Reliab R32(2):134–139
4. Bryant RE (1986) Graph based algorithms for Boolean function manipulation. IEEE Trans Comput c-35(8):677–691
5. Meinel C, Theobald T (1998) Algorithms and data structures in VLSI design: OBDD-foundations and applications. Springer, New York
6. Zhong J, Tong J, He Z (2010) An approach to use BDD during the fault tree editing and analyzing. In: Proceedings of the eight international conference on probabilistic safety assessment and management (PSAM)
7. Towhidi F, Lashkari A, Hosseini R (2009) Binary decision diagram (BDD). In: International conference on future computer and communication

8. Andersen HR (1999) An introduction to binary decision diagrams. Lecture notes for efficient algorithms and programs, IT University of Copenhagen
9. Lafferty J, Vardy A (1998) Ordered binary decision diagrams and minimal trellises CMU-CS-98-162. School of Computer Science Carnegie Mellon University, Pittsburgh
10. Vesely W, Dugan J, Fragola J et al (2002) Fault tree handbook with aerospace applications. National Aeronautics and Space Administration
11. Way YS, Hsia DY (2000) A simple component-connection method for building binary decision diagrams encoding a fault tree. Rel Eng Syst Saf 70:59–70
12. Bartlett LM, Andrews JD (2001) An ordering heuristic to develop the binary decision diagram based on structural importance. Rel Eng Syst Saf 72:31–38
13. Reay KA, Andrews JD (2002) A fault tree analysis strategy using binary decision diagrams. Rel Eng Syst Saf 78:45–56
14. Bollig B, Wegener I (1999) Complexity theoretical results on partitioned (nondeterministic) binary decision diagrams. Theory Comp Syst 32:487–503
15. Bishop CM (1995) Neural networks for pattern recognition. Clarendon, Oxford
16. Lhotak O (2006) Program analysis using binary decision diagrams. PhD thesis, McGill University
17. Lind-Nielsen J (2004) BuDDy: a binary decision diagram package
18. Whaley J (2007) Context-sensitive pointer analysis using binary decision diagrams. PhD thesis, Stanford University
19. Dutuit Y, Rauzy A (1999) A guided tour of minimal cutsets handling by means of binary decision diagrams. In: Proceedings of the probabilistic safety assessment conference, PSA'99, ANS, pp 55–62
20. Nusbaumer OPM (2007) Analytical solutions of linked fault tree probabilistic risk assessments using binary decision diagrams with emphasis on nuclear safety applications, Swiss Federal Institute of Technology Zurich

Chapter 8
Markov Processes

Everything existing in the universe is the fruit of chance
Democritus

8.1 Introduction

A Markov process is a continuous stochastic process in which future states are conditional only on the present state and are independent of previous states [1–4], e.g., a random time-varying process in which future states may be predicted only using the current state as an input [5–8]. This property of Markov processes is known as the Markov property. A Markov chain is a type of Markov process in which there are number of finite states (*S1, S2, S3….Sn*) that the process may exist at any given time. The probability of the process moving from *Si* to *Sj* is denoted by the transition probability *Pij* and the probability of the process remaining in the same state is denoted by the probability *Pii*.

Markov processes can be used to analyze the availability, reliability, and maintainability of systems [9–19]. A system is made of a number of components n, each of which at any given time may be operating successfully or not. The successful operation of the entire system depends on the operation or failure of its components. Therefore, the system may exist in one of two states.

- An operating state, where the system is operational even if some of its components have failed. A fully operational system is the one in which no components have failed.
- A failed state, where the system is not operational because of the failure of one or more of its components.

Such modeling provides a clear representation of all the states of a system as well as the transition between these states [4]. The failure of individual components in a system is also readily modeled using this method. One disadvantage, however, is that for large systems with many components, it is difficult to draw a diagram [6]. This is because for a system of n components, each with a failed or operating state, the number of states that exist is equal to 2^n.

DOI: 10.1007/978-0-85729-688-7_8,

8.2 Systems Availability Analyses

A step-by-step approach of the state–space method is demonstrated through the following example of a system with two repairable components.

8.2.1 Step 1: Constructing the Diagram

Figure 8.1 shows a diagram for the two component system. There are four states:

- State 1 (00): both components operational
- State 2 (10): component 1 failed and component 2 operational
- State 3 (01): component 2 failed and component 1 operational
- State 4 (11): both components failed

The failure rate (λ) and repair rate (μ) for each component are also represented in the diagram.

8.2.2 Step 2: Constructing the Transition Matrix

A transition matrix is created using the diagram. For a system of n components, the transition matrix will have the dimensions $n \times n$. The matrix is created by observing the changes between states and entering either the failure or repair rate causing the transition into the transition matrix. For example, for a transition between state i and state j (with $i \neq j$), the transition rate is entered into the ith row and jth column of the matrix. The diagonal elements of the matrix should be equal to 1 minus the sum of the other elements on the row. Any other element should be zero. The transition matrix is shown below.

$$T = \begin{vmatrix} 1-\lambda_1-\lambda_2 & \lambda_1 & \lambda_2 & 0 \\ \mu_1 & 1-\mu_1-\lambda_2 & 0 & \lambda_2 \\ \mu_2 & 0 & 1-\lambda_1-\mu_2 & \lambda_1 \\ 0 & \mu_2 & \mu_1 & 1-\mu_1-\mu_2 \end{vmatrix} \tag{8.1}$$

8.2.3 Step 3: Applying Markov Approach

The limiting state probability does not change in the further transition process. Therefore, this is expressed mathematically:

$$PT = P \tag{8.2}$$

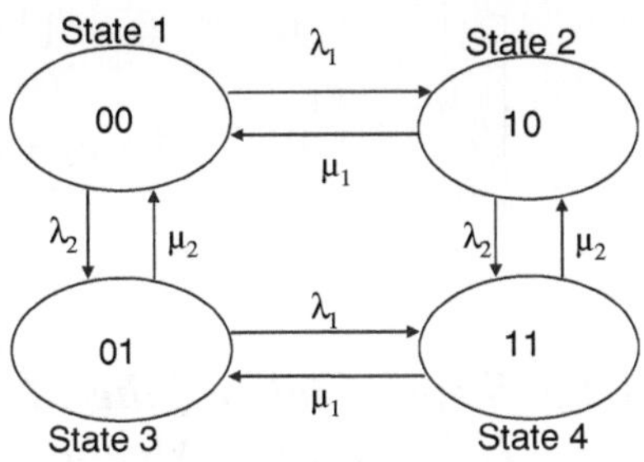

Fig. 8.1 Diagram for a two-component repairable system for availability analyses

where P is the limiting state probability vector and T is the transition matrix. This may be rewritten as:

$$P(T-I)=0 \tag{8.3}$$

where I is the identity matrix.

Substituting the transition matrix T into the above equation gives the following expression.

$$[P_1\ P_2\ P_3\ P_4]\times\begin{bmatrix}-(\lambda_1+\lambda_2) & \lambda_1 & \lambda_2 & 0\\ \mu_1 & -(\mu_1+\lambda_2) & 0 & \lambda_2\\ \mu_2 & 0 & -(\mu_2+\lambda_1) & \lambda_1\\ 0 & \mu_2 & \mu_1 & -(\mu_1+\mu_2)\end{bmatrix}=[0\ 0\ 0\ 0] \tag{8.4}$$

By taking the transpose of the equation, the general form of the equation is obtained.

$$\begin{bmatrix}-(\lambda_1+\lambda_2) & \mu_1 & \mu_2 & 0\\ \lambda_1 & -(\mu_1+\lambda_2) & 0 & \mu_2\\ \lambda_2 & 0 & -(\mu_2+\lambda_1) & \mu_1\\ 0 & \lambda_2 & \lambda_1 & -(\mu_1+\mu_2)\end{bmatrix}\begin{bmatrix}P_1\\P_2\\P_3\\P_4\end{bmatrix}=\begin{bmatrix}0\\0\\0\\0\end{bmatrix} \tag{8.5}$$

8.2.4 Step 4: Full Probability Condition

The sum of all the individual probabilities is equal to 1.

$$[P_1+P_2+P_3+P_4]=1$$

This condition is required to be able to solve the above equation as it contains only $n-1$ independent equations and there are four state variables involved. Therefore, any row within the above equation can be replaced with this condition, such as the first row.

$$\begin{bmatrix} 1 & 1 & 1 & 1 \\ \lambda_1 & -(\mu_1+\lambda_2) & 0 & \mu_2 \\ \lambda_2 & 0 & -(\mu_2+\lambda_1) & \mu_1 \\ 0 & \lambda_2 & \lambda_1 & -(\mu_1+\mu_2) \end{bmatrix} \begin{bmatrix} P_1 \\ P_2 \\ P_3 \\ P_4 \end{bmatrix} = \begin{bmatrix} 1 \\ 0 \\ 0 \\ 0 \end{bmatrix} \quad (8.6)$$

8.2.5 Step 5: Solving the Markov Matrix Equation Using Linear Algebra

As equation now contains four independent equations, it may be solved using linear algebra. In general, this yields the following expressions.

$$P_1 = \frac{\mu_1\mu_2}{(\mu_1+\lambda_1)(\mu_2+\lambda_2)} \quad (8.7)$$

$$P_2 = \frac{\lambda_1\mu_2}{(\mu_1+\lambda_1)(\mu_2+\lambda_2)} \quad (8.8)$$

$$P_3 = \frac{\mu_1\lambda_2}{(\mu_1+\lambda_1)(\mu_2+\lambda_2)} \quad (8.9)$$

$$P_4 = \frac{\lambda_1\lambda_2}{(\mu_1+\lambda_1)(\mu_2+\lambda_2)} \quad (8.10)$$

8.3 Example of Markov Chains for Reliability Analyses

A simple system with one active component named primary system and one standby redundant component named secondary system is connected with another component, i.e., switch [7]. Figure 8.2 shows this system.

When switch fails, it is unable to switch. The failure of the switch matters, if it is required to switch from the primary to the secondary system. If the switch fails after the standby spare is already in use, then the system can continue operation. However, if the switch fails before the primary unit fails, then the secondary system cannot be turned on and the system fails when the primary unit fails. The order in which the primary system and switch fail determines whether the system continues operation.

Figure 8.3 shows the Markov model for the secondary system. The *circles* represent states. The *arrows* represent events that cause transitions between states. *Arrows* are labeled with the rate at which the transition occurs, which is usually the failure rate of the failed component. The failure rate of the primary system when activated is λ_p. The failure rate of the secondary system when activated is the same and it is λ_p. The failure rate of the switch is λ_s. In the initial state or state 1

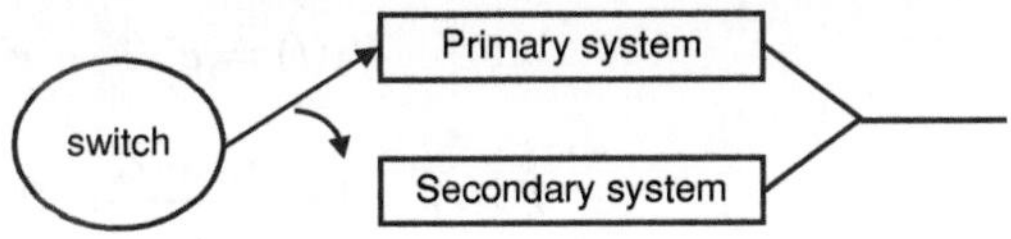

Fig. 8.2 Example system

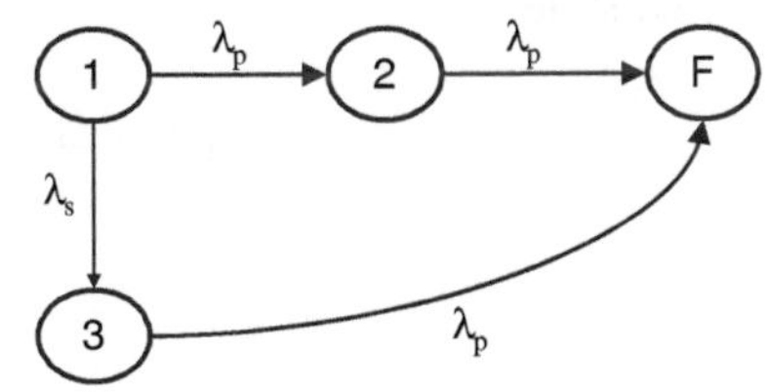

Fig. 8.3 Markov model for the secondary system

(Fig. 8.3), the primary system, the secondary system, and the switch are all functional. The two transitions from state 1 exist. The transition to state 2 represents the failure of the primary system. The transition to state 3 represents the failure of the switch. In state 2, the secondary system has been switched on, and a later failure of the switch is inconsequential. From state 3, failure of the secondary system leads to system failure, i.e., state *F* in the figure.

A Markov chain is quantified with a set of differential equations, with one equation for each state. Associated with each state is a state variable representing the time-dependent probability that the system is in that state. The states are assumed to be mutually exclusive. The set of differential equations associated with the Markov chain shown in Fig. 8.3 is given by equations.

$$\frac{d}{dt}P_1(t) = -(\lambda_p + \lambda_s)P_1(t) \tag{8.11}$$

$$\frac{d}{dt}P_2(t) = \lambda_p P_1(t) - \lambda_p P_2(t) \tag{8.12}$$

$$\frac{d}{dt}P_3(t) = \lambda_s P_1(t) - \lambda_p P_3(t) \tag{8.13}$$

$$\frac{d}{dt}P_F(t) = \lambda_p P_2(t) + \lambda_p P_3(t) \tag{8.14}$$

It is assumed that the initial state at time zero is state 1. $P_1(0) = 1$ and $P_i(t) = 0; i \neq 1$.

The solution of the differential equations is given by equations.

$$P_1(t) = e^{-(\lambda_S+\lambda_P)t} \tag{8.15}$$

$$P_2(t) = \frac{\lambda_P}{\lambda_S}\left(e^{-\lambda_P t} - e^{-(\lambda_P+\lambda_s)t}\right) \tag{8.16}$$

$$P_3(t) = e^{-\lambda_{\rm p}t} - e^{-(\lambda_{\rm p}+\lambda_{\rm s})t} \tag{8.17}$$

$$P_F(t) = 1 - \frac{\lambda_{\rm p}+\lambda_{\rm s}}{\lambda_{\rm s}} e^{-\lambda_{\rm p}t} + \frac{\lambda_{\rm p}}{\lambda_{\rm s}} e^{-(\lambda_{\rm p}+\lambda_{\rm s})t} \tag{8.18}$$

The reliability of the standby system is the probability that the system is not in a failed state.

$$R(t) = \frac{\lambda_{\rm p}+\lambda_{\rm s}}{\lambda_{\rm s}} e^{-\lambda_{\rm p}t} + \frac{\lambda_{\rm p}}{\lambda_{\rm s}} e^{-(\lambda_{\rm p}+\lambda_{\rm s})t} \tag{8.19}$$

References

1. Markov AA (1954) Theory of algorithms (Teoriya algoritmov). Academy of Sciences of the USSR
2. Howard RA (1971) Dynamic probabilistic systems. Wiley, New York
3. Kemeny JG, Snell JL, Thompson GL (1974) Introduction to finite mathematics. Prentice-Hall, Englewood Cliffs, NJ
4. Grinstead C, Snell J (2003) Grinstead and Snell's introduction to probability, 2nd edn. American Mathematical Society
5. Villemeur A (1992) Reliability, availability, maintainability and safety assessment: methods and techniques. Wiley, New York
6. Li W (2005) Risk assessment of power systems: models, methods, and applications. Wiley, IEEE, New York
7. Vesely W, Dugan J, Fragola J et al. (2002) Fault tree handbook with aerospace applications. National Aeronautics and Space Administration, NASA
8. Pukite J, Pukite P (1998) Modeling for reliability analysis. IEEE, New York
9. Doeblin W (1937) Exposé de la Théorie des Chaines Simple Constantes de Markov à un Nombre Fini d'Etats. Rev Mach de l'Union Interbalkanique 2:77–105
10. Castelo R, Perlman MD (2002) Learning essential graph Markov models from data. Technical report no. 416
11. Litterman RB (1983) A random walk, Markov model for the distribution of time series. J Business Econ Stat 1(2):169–173
12. IEC 1165 (1995) Application of Markov techniques. IEC
13. MIL-HDBK-338B (1998) Electronic reliability design handbook. DoD
14. Xie M, Da Y, Poh K (2004) Computing system reliability: models and analysis. Kluwer Academic, New York
15. Smith DJ (2001) Reliability maintainability and risk. Butterworth-Heinemann, Woburn, MA
16. Dhillon BS (2007) Applied reliability and quality: fundamentals, methods and procedures. Springer, London
17. Levitin G (2007) Computational intelligence in reliability engineering, evolutionary techniques in reliability analysis and optimization. Springer, Berlin Heidelberg
18. Bertsekas DP, Tsitsiklis JN (2002) Introduction to probability. Athena Scientific, Belmont, MA
19. Anders GJ (1989) Probability concepts in electric power systems. Wiley, New York

Chapter 9
Reliability Block Diagram

If something can go wrong, it will
Captain Ed Murphy (1949)

9.1 Introduction

The reliability block diagram is an inductive method used to analyze systems and assess their reliability [1–4]. The reliability block diagram involves representing a system and its distinctive components by using a graphical representation that can be used to analyze the probability of system failure [5–15]. The blocks within the block diagram are linked depending on their effects on the system. The blocks represent the smallest entities of the system, which are not further divided, i.e., components of the system or the blocks represent groups of components.

9.2 Components in Series

If the individual components of a system are connected in series, the failure of any component causes the system to fail. Therefore, if one block fails, the system fails. If the individual components of a system are connected in parallel, the failures of all components cause the system to fail. Therefore, if all blocks fail, the system fails.

Figure 9.1 shows a reliability block diagram of a system containing n components connected in series. If E_i is the event that component C_i is operating at time t, then the reliability of the system may be written as:

$$R = P[E_1 * E_2, \ldots, E_i, \ldots, E_n] \quad (9.1)$$

If all the events are independent of each other, then the reliability may be expressed as:

$$R = \prod_{i=1}^{n} P[E_i] \quad (9.2)$$

M. Čepin, *Assessment of Power System Reliability*,
DOI: 10.1007/978-0-85729-688-7_9, © Springer-Verlag London Limited 2011

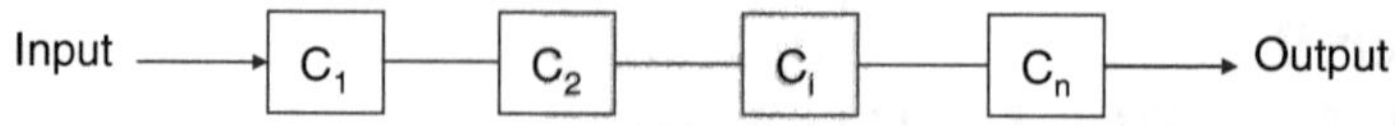

Fig. 9.1 Reliability block diagram with n components in series

Table 9.1 Examples of calculations: reliability of series system

Reliability of component 1: $R1$	Reliability of component 2: $R2$	Reliability of component 3: $R3$	Reliability of the system: R
0.9	0.9	0.9	0.729
0.8	0.8	0.8	0.512
0.8	0.9	0.99	0.7128

Or in other words, the reliability of the system R is calculated as a product of the reliabilities of its components R_i:

$$R = \prod_{i=1}^{n} R_i \tag{9.3}$$

Table 9.1 shows examples of calculations in the case of a system consisting of three components in series (e.g., valve 1, pump, and valve 2), where the reliabilities of components are known.

9.3 Parallel Components

Figure 9.2 shows a reliability block diagram of a system containing n parallel components. If E_i is the event that component C_i is operating at time t, then the reliability of the system may be written as:

$$R = P[E_1 + E_2 + \cdots + E_i + \cdots + E_n] \tag{9.4}$$

$\overline{E_i}$ is the complement of the event E_i and represents the event that component i has failed at time t. Therefore, the reliability of the system may be expressed as

$$R = 1 - P[\overline{E_1 + E_2 + \cdots + E_i + \cdots + E_n}] \tag{9.5}$$

$$R = 1 - P[\overline{E_1}, \overline{E_2}, \ldots, \overline{E_i}, \ldots, \overline{E_n}] \tag{9.6}$$

If all the events are independent, then the reliability may be written as:

$$R = 1 - \prod_{i=1}^{n} P[\overline{E_i}] \tag{9.7}$$

Or in other words, the reliability of the system is calculated from the reliabilities of its components:

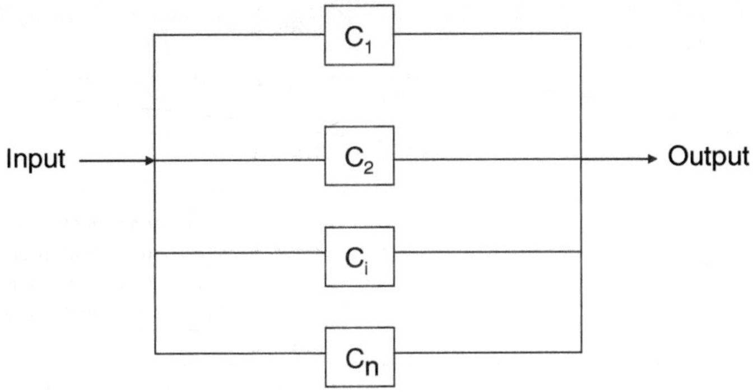

Fig. 9.2 Reliability block diagram with n parallel components

Table 9.2 Examples of calculations: reliability of parallel system

Reliability of component 1: $R1$	Reliability of component 2: $R2$	Reliability of component 3: $R3$	Reliability of the System: R
0.9	0.9	0.9	0.999
0.8	0.8	0.8	0.992
0.8	0.9	0.99	0.9998

$$R = 1 - \prod_{i=1}^{n} (1 - R_i) \tag{9.8}$$

Table 9.2 shows examples of calculations in the case of a system consisting of three components in parallel (e.g., three redundant pumps of the system and operation of one is enough to fulfill the success criteria of the system), where the reliabilities of components are known.

Figure 9.3 shows that the system reliability increases fast with increased number of parallel components. The redundant components largely increase the reliability of the system. Figure 9.4 shows that the system reliability increases fast with increased reliability of its components.

Figure 9.5 is an example system of seven components. Quantification of system reliability goes in steps hand in hand with system simplification presented within the reliability block diagram. The four components in series C_{a1}, C_{a2}, C_{a3}, and C_{a4} and three components in series C_{b1}, C_{b2}, and C_{b3} are represented by one group each in the first quantification step getting the C_a and C_b, respectively.

The corresponding reliabilities of groups a and b are R_a and R_b.

$$R_a = \prod_{i=1}^{4} R_i = R_{a1} \times R_{a2} \times R_{a3} \times R_{a4} \tag{9.9}$$

$$R_b = \prod_{i=1}^{3} R_i = R_{b1} \times R_{b2} \times R_{b3} \tag{9.10}$$

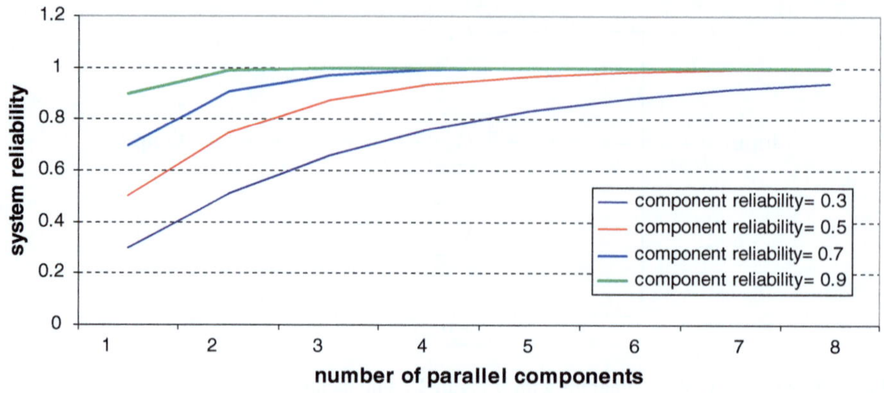

Fig. 9.3 System reliability versus number of parallel components

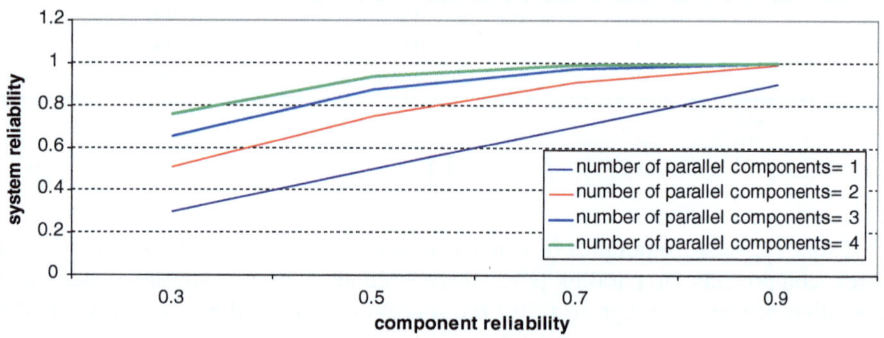

Fig. 9.4 System reliability versus number of parallel components

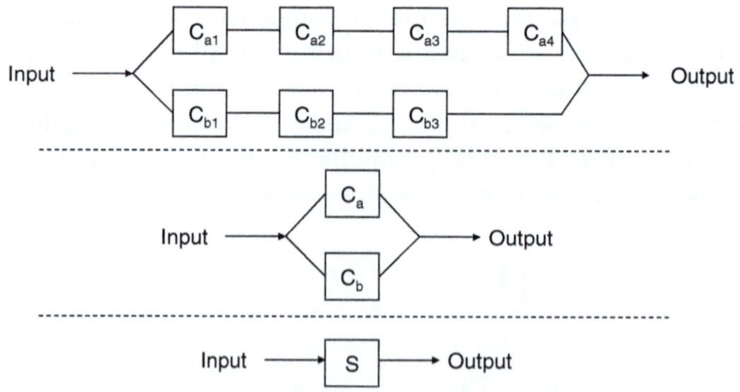

Fig. 9.5 Example system and its simplification steps

The system reliability is calculated in the second step, where the parallel groups a and b are joined as parallel components into the system represented by S. The reliability of the system S is calculated from its parallel groups a and b.

$$R = 1 - \prod_{i=1}^{2}(1 - R_i) = 1 - (1 - R_a)(1 - R_b) \tag{9.11}$$

Going backward and considering the previous steps, one may express the system reliability with the reliabilities of the components instead of the reliabilities of component groups.

$$R = 1 - \prod_{i=1}^{2}(1 - R_i) = 1 - (1 - R_{a1} \times R_{a2} \times R_{a3} \times R_{a4})(1 - R_{b1} \times R_{b2} \times R_{b3}) \tag{9.12}$$

References

1. Villemeur A (1992) Reliability, availability, maintainability and safety assessment: methods and techniques. Wiley, New York
2. Kumamoto H, Henley EJ (1996) Probabilistic risk assessment and management for engineers and scientists. IEEE, New York
3. Billinton R, Allan R (1996) Reliability evaluation of power systems. Plenum, New York
4. Vesely W, Dugan J, Fragola J et al (2002) Fault tree handbook with aerospace applications. National Aeronautics and Space Administration
5. Birolini A (2007) Reliability engineering theory and practice. Springer, Berlin, Heidelberg
6. Lewis EE (1996) Introduction to reliability engineering. Wiley, New York
7. MIL-HDBK-338B (1998) Electronic Reliability Design Handbook, DoD
8. Smith DJ (2001) Reliability, maintainability, and risk. Butterworth-Heinemann, Woburn, MA
9. O'Connor PDT (1991) Practical reliability engineering. Wiley, Chichester
10. Aven T (1992) Reliability and risk analysis. Elsevier, London
11. Anders GJ (1989) Probability concepts in electric power systems. Wiley, New York
12. Levitin G (2007) Computational intelligence in reliability engineering evolutionary techniques in reliability analysis and optimization. Springer, Berlin, Heidelberg
13. Chapman D (1997) Electrical design: a good practice guide. CDA Publication 123
14. Chambal SP, Keats JB (2000) Evaluating complex system reliability using reliability block diagram simulation when little or no failure data are available. Qual Eng 13(2):169–177
15. Wang W, Loman JM, Arno RG et al (2004) Reliability block diagram simulation techniques applied to the IEEE Std 493 standard network. IEEE Trans Ind Appl 40(3):887–895

Chapter 10
Common Cause Failures

The most common cause of insufficient results is insufficient action

Brian Koslow

10.1 Introduction

Common cause failure (CCF) events are a subset of dependent events in which two or more component fault states exist at the same time and are a direct result of a shared root cause [1–4]. The definition of common cause failure is closely related to the general definition of dependent failure. The events A and B are dependent if:

$$P(A \cap B) \neq P(A) \times P(B) \tag{10.1}$$

where $P(x)$ is the probability of event x.

If the dependency exists between parallel events, the probability of system failure is larger than the product of failure probabilities of all parallel events:

$$P(A \cap B) > P(A) \times P(B) \tag{10.2}$$

Common cause failure results from the coexistence of two main factors: (i) a susceptibility for components to fail or become unavailable because of a particular root cause of failure and (ii) a coupling mechanism that creates the condition for multiple components to be affected by the same cause. An example is the case where two parallel relief valves fail to open at the required pressure because of set points being set too high, as a result of incorrect procedure. What makes the two valves fail together is a common calibration procedure and perhaps the maintaining of common maintenance personnel. These commonalities are the coupling mechanism of the failure event in this case.

Another example is the case where the components A and B fail to function. The root cause for the failures is the high humidity, because the components are susceptible to high humidity. The coupling mechanism is the location of both components in the same room. Figure 10.1 shows elements of dependent events in general.

Procedure of common cause failure analysis is commonly organized into two main phases:

DOI: 10.1007/978-0-85729-688-7_10,

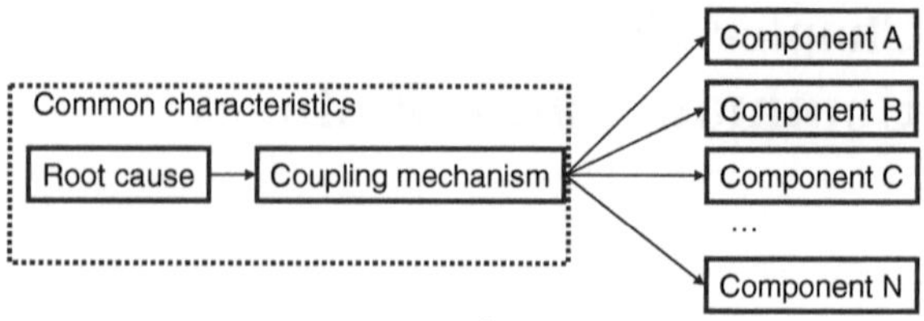

Fig. 10.1 Elements of dependent events

Phase I Identification phase
Phase II Evaluation phase

The main objective of the identification phase is to identify those groups of components within the system whose common cause failures contribute significantly to the system failure probability or to the system unavailability. The main objectives of the evaluation phase are: to select the method for evaluation of common cause failures, to obtain quantitative data about parameters needed for evaluation, and to perform qualitative and quantitative evaluation of the common cause failures.

10.2 Identification Phase

The prerequisite for the identification phase is the system familiarization, where the functions of the system are acknowledged and components of the system and their interactions are studied [5–8]. The boundaries of the system and the resolution of modeling need to be defined. The definition of boundaries means that it is determined what is considered within the system and what is considered out of the system. The resolution of modeling means that the smallest items or components of the analysis are defined. For example, diesel generator of the system under consideration can be considered as a component, which is not further divided. Or alternatively, it can be decided that specific parts of diesel generator are considered as components of the system.

Very helpful to the identification phase is the reliability model of the system, which can be done by the system logic model, e.g., fault tree or by the block diagram [8, 9].

Then, the groups of components within the system whose common cause failures could contribute to the system failure probability or to the system unavailability are identified.

Figure 10.2 shows a simple example of three parallel pumps in one system, which have to deliver water to the tank for 4 h. The capacity of one pump fulfils the success criteria of the system, so three pumps have to fail for the system failure. The identification of one group of components susceptible to common cause failure is performed for the simple example. The group of components includes all three pumps. The root causes for the failures are the high humidity and high temperatures. The coupling mechanism is the location of the components in

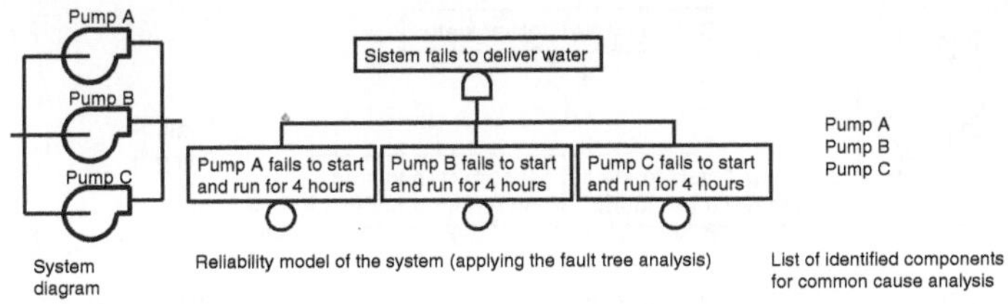

Fig. 10.2 Three components susceptible to common cause failures (fault tree)

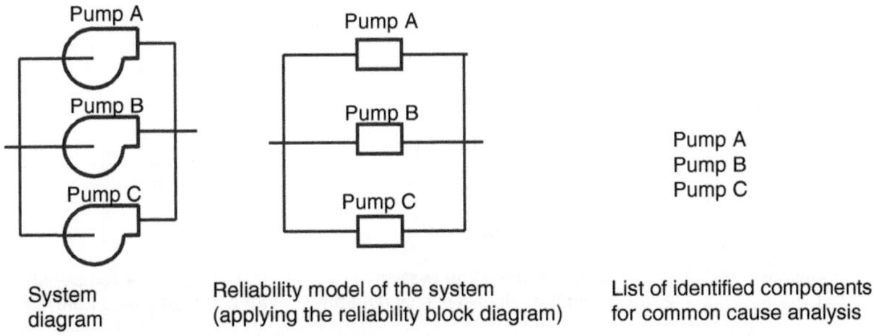

Fig. 10.3 Three components susceptible to common cause failures (reliability block diagram)

the same room. Figure 10.3 shows the same example for the case if reliability block diagram is the reliability method instead of the fault tree.

Figure 10.4 shows a similar example, which is even simpler as it consists of two parallel pumps in one system, which have to deliver water to the tank for 4 h. The capacity of one pump fulfils the success criteria of the system, so both pumps have to fail for the system failure. The identification of one group of components susceptible to common cause failure is performed for the simple example. The group of components includes both pumps. The root causes and the coupling mechanism are the same as at previous example with three pumps. Figure 10.5 shows the same example of two pumps for the case if reliability block diagram is the reliability method instead of the fault tree.

The coupling mechanism can exist when two or more components exhibit similar characteristics, both in the cause and in the failure mechanism. The focus should be placed on considering the components of the system, which share one or more common characteristics such as:

- Same design
- Same manufacturer
- Same or similar function
- Same personnel dealing with the operation or with maintenance or with installation and construction
- Same procedures or instructions
- Same interface between components or systems

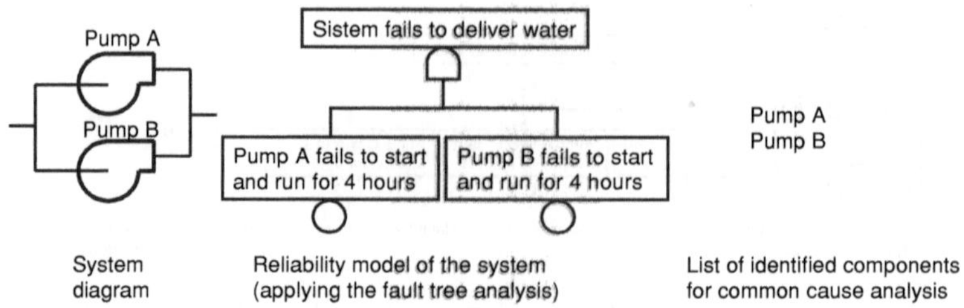

Fig. 10.4 Two components susceptible to common cause failures (fault tree)

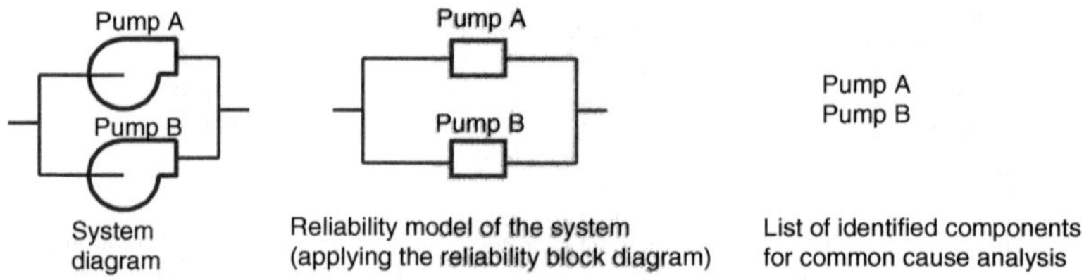

Fig. 10.5 Two components susceptible to common cause failures (reliability block diagram)

- Same location
- Same environment

10.3 Evaluation Phase

The evaluation phase includes selection of the method for evaluation of common cause failures, preparation of quantitative data about parameters needed for evaluation, and qualitative and quantitative evaluation of the common cause failures.

A detailed qualitative and quantitative analysis can be achieved through the following steps:

- Selection of the method for evaluation of common cause failures
- Incorporation of common cause failure events into the system reliability model
- Parameter estimation
- System quantification
- Result evaluation and documentation

10.3.1 Selection of the Method for Evaluation of Common Cause Failures

The methods for evaluation of common cause failures include the following methods:

- Beta factor method
- Basic parameter method
- Multiple Greek letter method
- Alpha factor method

The full effect of the difference between the methods is observed for the systems with more than two or three parallel branches or portions or redundant lines. If there are only two parallel components in the system, the differences between the methods are not notable in the whole context. The simplest method for evaluation of common cause failures is the beta factor method.

10.3.1.1 Beta Factor Method

The beta factor method is a method where the likelihood of the common cause failure is evaluated in relation to the random failure probability of the components susceptible to a common cause failure.

If the components failures are not completely independent from others, the original random failure probability of failure mode of specific component can be divided into two parts: independent failure probability and common cause failure probability. The beta factor method is one parameter method, where the factor β directs the extent of common cause failure probability related to the original random failure probability.

According to the beta factor method, a part of the component failure rate or component failure probability can be associated with common cause failure events shared by the other components in that group. According to this method, whenever a common cause failure event occurs, all components within the common cause component group are assumed to fail.

The failure events of the system of two parallel pumps are the following:

A_I Single independent failure of pump *A* (basic event *A*)
C_{AB} Failure of pumps A and B from common causes
B_I Single independent failure of pump B (basic event B)

Component *A* fails if any of the events occur: single independent failure or common cause failure. The equivalent Boolean representation of the total failure of pump *A* is:

$$A_T = A_I + C_{AB} \tag{10.3}$$

Similarly, the equivalent Boolean expression for the total failure of pump B is:

$$B_T = B_I + C_{AB} \tag{10.4}$$

In terms of probabilities, the failure probability of pump A is:

$$P_A = P_{AI} + P_{\text{CCF}} = P_{AI} + P_{AB} \tag{10.5}$$

where

$$P_{AI} = (1 - \beta) \times P_A \tag{10.6}$$

$$P_{\text{CCF}} = P_{AB} = \beta \times P_A \tag{10.7}$$

In terms of probabilities, the failure probability of pump B is:

$$P_B = P_{BI} + P_{\text{CCF}} = P_{BI} + P_{AB} \tag{10.8}$$

where

$$P_{BI} = (1 - \beta) \times P_B \tag{10.9}$$

$$P_{\text{CCF}} = P_{AB} = \beta \times P_B \tag{10.10}$$

The equivalent Boolean expression for a system failure is written from components expressions and derived applying Boolean rules ($C_{AB} \times C_{AB} = C_{AB}$, $C_{AB} \times A_I + C_{AB} = C_{AB}$):

$$\begin{aligned} S &= (A_I + C_{AB}) \times (B_I + C_{AB}) = A_I \times B_I + C_{AB} \times B_I + A_I \times C_{AB} + C_{AB} \times C_{AB} \\ &= A_I \times B_I + C_{AB} \end{aligned} \tag{10.11}$$

If both pumps in the system of two parallel pumps are independent, the failure probability of the system (P_S) is the product of the failure probabilities of pumps:

$$P_S = P_A \times P_B \tag{10.12}$$

If the pumps share the cause of their failure mode (e.g., pump fails to start or, e.g., pump fails to run for 4 h), the failure probability of the system is calculated differently.

$$P_S = (1 - \beta) \times P_A \times (1 - \beta) \times P_B + P_{\text{CCF}} \tag{10.13}$$

where

$$P_{\text{CCF}} = \beta \times P_A \tag{10.14}$$

The β factor is obtained through the data base, which can mostly base on historical data by determining the percentage of all the component failures in which multiple similar components failed versus single components failures. If β factor is not known, a general value of 0.1 is sometimes used.

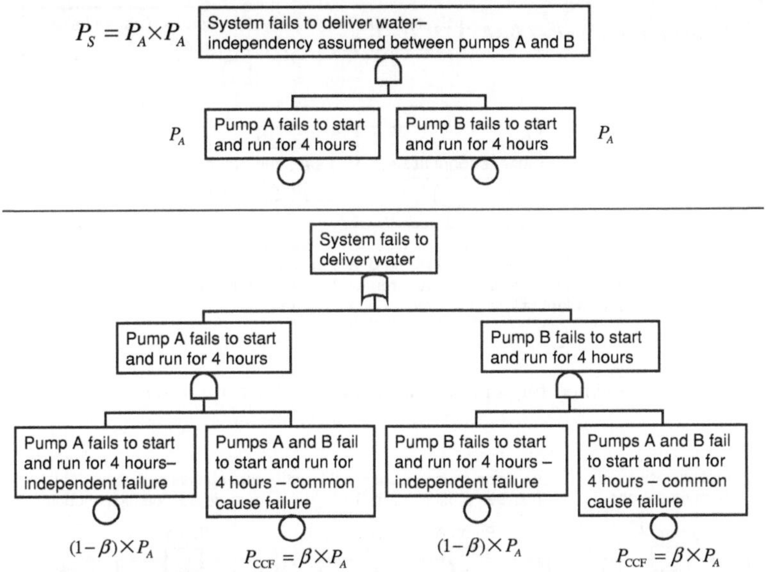

Fig. 10.6 Beta factor method for common cause failures (fault tree)

Because the redundant pumps are usually similar, their failure probabilities may be identical and the expression can be simplified.

$$P_S = P_A \times P_A \tag{10.15}$$

For the same failure probabilities for both components A and B, the full expression for the system failure probability is the following.

$$P_S = (1 - \beta) \times P_A \times (1 - \beta) \times P_A + \beta \times P_A \tag{10.16}$$

Figure 10.6 shows evolution of the change from fault tree model without consideration of common cause failures to a model with consideration of common cause failures. Figure 10.7 shows a different fault tree of the system, but it is identical to the previous version.

Figure 10.8 shows the case if the reliability block diagram is considered instead of the fault tree for the quantification of the system failure probability. If the independent pumps are considered, there are two system components on the *left* of Fig. 10.8. If the common cause is considered with the beta factor method, the *middle* figure is developed from the *left* figure. The abstract components CCF are added to the diagram to represent the common cause failure contribution. Finally, on the *right side* of the figure, one can see different graphical consideration of the common cause failure contribution, which is quantitatively identical to the middle part of the figure.

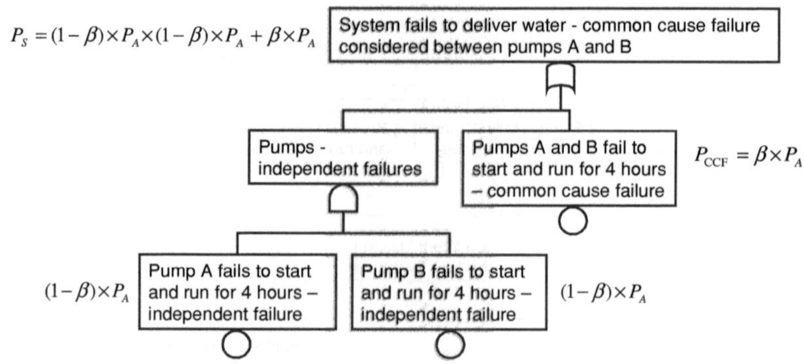

Fig. 10.7 Beta factor method for common cause failures (fault tree variant)

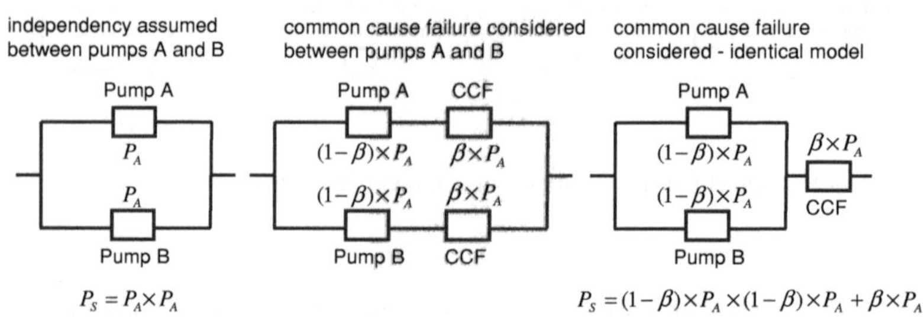

Fig. 10.8 Beta factor method for common cause failures (reliability block diagram)

If there are more parallel components in the system, the calculation is a little more complex. In a system of three redundant components *A*, *B*, and *C*, the common cause failure events are C_{AB}, C_{AC}, C_{BC}, and C_{ABC}.

The failure events of the system of three parallel components are four for each of parallel components and are the following:

A_I Single independent failure of component *A* (basic event *A*)
C_{AB} Failure of components *A* and *B* (and not *C*) from common causes
C_{AC} Failure of components *A* and *C* (and not *B*) from common causes
C_{ABC} Failure of components *A*, *B*, and *C* from common causes
B_I Single independent failure of component *B* (basic event *B*)
C_{AB} Failure of components *A* and *B* (and not *C*) from common causes
C_{BC} Failure of components *B* and *C* (and not *A*) from common causes
C_{ABC} Failure of components *A*, *B*, and *C* from common causes
C_I Single independent failure of component *C* (basic event *C*)
C_{BC} Failure of components *B* and *C* (and not *A*) from common causes
C_{AC} Failure of components *A* and *C* (and not *B*) from common causes
C_{ABC} Failure of components *A*, *B*, and *C* from common causes

Some of them repeat, but all failure events related to each of parallel components are listed intentionally.

Component A fails if any of the related four events occur. The equivalent Boolean representation of total failure of component A is:

$$A_T = A_I + C_{AB} + C_{AC} + C_{ABC} \tag{10.17}$$

Similarly, the equivalent Boolean expression for a total failure of component B is:

$$B_T = B_I + C_{AB} + C_{BC} + C_{ABC} \tag{10.18}$$

Similarly, the equivalent Boolean expression for a total failure of component C is:

$$C_T = C_I + C_{AC} + C_{BC} + C_{ABC} \tag{10.19}$$

In terms of probabilities, the failure probability of component A is:

$$P_A = P_{AI} + P_{\mathrm{CCF}} = P_{AI} + P_{ABC} \tag{10.20}$$

where

$$P_{AI} = (1 - \beta) \times P_A \tag{10.21}$$

$$P_{\mathrm{CCF}} = P_{ABC} = \beta \times P_A \tag{10.22}$$

In terms of probabilities, the failure probability of component B is:

$$P_B = P_{BI} + P_{\mathrm{CCF}} = P_{BI} + P_{ABC} \tag{10.23}$$

where

$$P_{BI} = (1 - \beta) \times P_B \tag{10.24}$$

$$P_{\mathrm{CCF}} = P_{ABC} = \beta \times P_B \tag{10.25}$$

In terms of probabilities, the failure probability of component C is:

$$P_C = P_{CI} + P_{\mathrm{CCF}} = P_{CI} + P_{ABC} \tag{10.26}$$

where

$$P_{CI} = (1 - \beta) \times P_C \tag{10.27}$$

$$P_{\mathrm{CCF}} = P_{ABC} = \beta \times P_C \tag{10.28}$$

The general expression for contributions of failure probabilities of m parallel components includes the following parameters:

P_1 Probability of independent part of component failure probability

P_2 Probability of common cause part of component failure probability considering common cause failure of two components

P_3 Probability of common cause part of component failure probability considering common cause failure of three components

For the case of four parallel components ($m = 4$), the probabilities P_1, P_2, P_3, and P_4 are calculated.

The general expression for contributions of failure probabilities of m parallel components for the component t is the following:

$$P_k = \begin{cases} (1-\beta) \times P_t; & k = 1 \\ 0; & 1 < k < m \\ \beta \times P_t; & k = m \end{cases} \tag{10.29}$$

This expression shows that only the independent part of the component failure and the common cause failure contribution of all components failures at the same time are considered in the beta factor method. The common cause contribution for the common cause failure of two components in the system of three parallel components equals zero. This expression shows that for three parallel components $t = \{A, B, C\}$, there are three equations for the independent part of the failure probability, one for each component. This expression implies that:

$$P_1 = (1-\beta) \times P_t \quad t = \{A, B, C\} \tag{10.30}$$

$$P_2 = 0 \quad \forall t \tag{10.31}$$

$$P_3 = P_m = \beta \times P_t \quad t = \{A, B, C\} \tag{10.32}$$

$$\beta = \frac{P_m}{P_1 + P_m} \tag{10.33}$$

The equivalent Boolean expression for a system failure is written from components expressions and derived applying Boolean rules:

$$\begin{aligned} S &= (A_I + C_{AB} + C_{AC} + C_{ABC}) \times (B_I + C_{AB} + C_{BC} + C_{ABC}) \\ &\quad \times (C_I + C_{AC} + C_{BC} + C_{ABC}) \\ &= A_I \times B_I \times C_I + A_I \times C_{BC} + B_I \times C_{AC} + C_I \times C_{AB} + C_{ABC} \end{aligned} \tag{10.34}$$

If all three parallel components in the system of three parallel components are independent, the failure probability of the system (P_S) is the product of the failure probabilities of components:

$$P_S = P_A \times P_B \times P_C \tag{10.35}$$

If the components share the cause of their failure mode, the failure probability of the system is calculated considering the rare event approximation:

$$P_S = P_A \times P_B \times P_C + P_{\text{CCF}} = P_A \times P_B \times P_C + P_{ABC} \tag{10.36}$$

The failure probability of the system can be calculated also through the general expression:

$$P_S = P_1^3 + P_3 \tag{10.37}$$

If another method is selected instead of the beta factor method, the basic equations, which represent the selected common cause failure method, replace the equations of the beta factor method.

10.3.1.2 Basic Parameter Method

The basic parameter model refers to the straightforward definition of the probabilities of the basic events P_k^m. The total failure probability (P_t) of a component in a common cause group of m components is:

$$P_t = \sum_{k=1}^{m} \binom{m-1}{k-1} \times P_k^m \tag{10.38}$$

where

$$\binom{m-1}{k-1} = \frac{(m-1)!}{(k-1)!(m-k)!} \tag{10.39}$$

where the events P_k^m and P_j^m are mutually exclusive for all k, j.

10.3.1.3 Multiple Greek Letter Method

The multiple Greek letter method is used for a more accurate analysis of systems with higher levels of redundancy. The multiple Greek letter method uses other parameters in addition to the β factor to distinguish among common cause events affecting different numbers of components in a higher order redundant system. The multiple Greek letter parameters consist of the total failure probability, which includes the effects of all independent and common cause contributions to that component failure, and a set of failure fractions, which are used to quantify the conditional probabilities of all the possible ways a common cause failure of a component can be shared with other components in the same group, given component failure has occurred. For a group of three components, three different parameters are defined:

$$P_k = \begin{cases} (1-\beta) \times P_t; & k = 1 \\ \frac{1}{2}\beta(1-\chi) \times P_t; & k = 2 \\ \beta \times \chi \times P_t; & k = 3 \end{cases} \tag{10.40}$$

The general expression for the multiple Greek letter method is the following:

$$P_k = \frac{1}{\binom{m-1}{k-1}}\left(\prod_{i=1}^{k}\rho_i\right) \times (1-\rho_{k+1}) \times P_t \quad \text{where} \quad \rho_1 = 1, \rho_2 = \beta, \rho_3 = \chi, \rho_{m+1} = 0 \tag{10.41}$$

The beta factor method is a special case of the multiple Greek letter method. If χ of the multiple Greek letter method equals one, this becomes the beta factor method.

10.3.1.4 Alpha Factor Method

The alpha factor method is a multiparameter method that can handle any redundancy level. The alpha factor method develops common cause failure probabilities from a set of failure ratios and the total component failure probabilities. The parameters of the model are:

P_t Total failure probability of each component because of all independent and common cause events

α_k Fraction of the total failure probability of events that occur in the system and involve the failure of k components because of a common cause

The general expression for the alpha factor method is the following:

$$P_k = \frac{k}{\binom{m-1}{k-1}}\frac{\alpha_k}{\alpha_t} \times P_t \quad \text{where} \quad \alpha_t = \sum_{k=1}^{m} k \times \alpha_k, \quad \text{where } k = 1, 2, \ldots, m \tag{10.42}$$

10.3.2 Incorporation of Common Cause Failure Events Into the System Reliability Model

Incorporation of common cause failure events into the system reliability model is a step, where the components susceptible to common cause failures are represented in the reliability model of the system in a way that enables their evaluation.

Figures 10.6 and 10.7 presented in previous sections show variants of the incorporation of common cause failure events into the system reliability model if the fault tree is used for the development and evaluation of the reliability of the system and the beta factor method is selected as a common cause analysis method.

10.3.3 Parameter Estimation

The parameter estimation is a step, where the quantitative estimation of the common cause parameters is made. The estimation is done based on the available information about the root causes and coupling mechanisms of each specific set of failures under investigation. Databases about common cause failures are helpful for determining the needed parameters [10–16].

If the beta factor method is selected for the common cause analysis of a particular system, the parameters β need to be determined for all determined groups of common cause failures: For example, β_{CB}—beta factor for circuit breakers that may share the same cause of failure, β_T—beta factor for transformers that may share the same cause of failure, and β_P—beta factor for pumps that may share the same cause of failure.

10.3.4 System Quantification

The system quantification step includes the quantification of the system reliability or system unreliability or system availability or system unavailability considering the common cause failures.

10.3.5 Result Evaluation and Documentation

The results evaluation and documentation step is important for making the conclusions of the analysis.

A comparison can be made with quantification without considering common cause failures. Generally, the contribution of common cause failures is higher in highly reliable systems where the redundancy is an important factor in system design [1–2, 17].

A comparison with proposed system modifications can be made in sense to determine which modification is the most effective in terms of increased reliability.

10.4 Example Quantification

The example system of three parallel components is evaluated with and without consideration of common cause failures. The parallel components are identical with the same failure probability of $P = 10^{-3}$. The system failure probability without consideration of common cause failures is calculated as:

$$P_S = P_A \times P_B \times P_C = 10^{-9} \tag{10.43}$$

The system failure probability with consideration of common cause failures with the beta factor method and with $\beta = 0.1$ is calculated as:

$$P_S = P_t^3 \times (1 - \beta)^3 + P_t \times \beta \approx 0.0001 \quad (10.44)$$

This probability is considerably higher as in the case without consideration of common cause failures.

References

1. NUREG/CR-4780 (1988) Procedures for treating common cause failures in safety and reliability studies. NRC
2. Mosleh A, Rasmuson DM, Marshall FM (1998) Guidelines on modeling common-cause failures in probabilistic risk assessment. NUREG/CR-5485, NRC
3. Fleming K, Mosleh A (1985) Common-cause data analysis and implications in system modeling. In: Proceedings of international topical meeting probabilistic safety methods and applications
4. Vaurio JK (1998) An implicit method for incorporating common-cause failures in system analysis. IEEE Trans Rel 47(2):173–180
5. Vaurio JK, Jankala KE (2002) Quantification of common cause failure rates and probabilities for standby-system fault trees using international event data sources. In: Proceedings of PSAM 6 conference, San Juan
6. PRA Procedures Guide (1983) NUREG/CR-2300, NRC
7. NUREG/CR-4550 (1990) Analysis of core damage frequency: internal events methodology. NRC
8. Roberts NH, Vesely WE, Haasl DF, Goldberg FF (1981) Fault tree handbook. NUREG-0492, NRC, Washington
9. Vesely W, Dugan J, Fragola J et al (2002) Fault tree handbook with aerospace applications. National Aeronautics and Space Administration
10. Common-Cause Failure Event Insights (2003) Emergency diesel generators, NUREG/CR-6819, vol 1. NRC
11. Common-Cause Failure Event Insights (2003) Motor-operated valves, NUREG/CR-6819, vol 2. NRC
12. Common-Cause Failure Event Insights (2003) Pumps, NUREG/CR-68 19, vol 3. NRC
13. Common-Cause Failure Event Insights (2003) Circuit breakers. NUREG/CR-6819, vol 4. NRC
14. IEEE-Std 500 (1984) IEEE guide to the collection and presentation of electrical, electronic, and sensing component reliability data for nuclear-power generating stations. IEEE
15. NEA/CSNI/R(2008)1 (2008) Collection and analysis of common-cause failures of switching devices and circuit breakers. NEA
16. Marshall FM, Rasmuson D, Mosleh A (1998) Common cause failure data collection and analysis system. NUREG/CR-6268, INEEL/EXT-97-00696, NRC
17. Čepin M (2010) Assessment of switchyard reliability with the fault tree analysis. In: Proceedings of the international conference on probabilistic safety assessment and management. IAPSAM

Part III
Power Flow Analysis

Chapter 11
Methods for Power Flow Analysis

Nearly all men can stand adversity, but if you want to test a man's character, give him power

Abraham Lincoln

11.1 Introduction

The methods for power flow analysis are the prerequisite for planning and designing the power systems and for their future expansion [1–5]. They are important for determining the optimal operation of existing systems [6–10].

Several methods have been developed and have been subjected to valuable improvements [11–18]. The most known methods include the Newton–Raphson method [3, 4, 9, 10], the Gauss–Seidel method [1, 2, 5, 9, 14], the fast decoupled load flow method [11, 19, 22–25], the direct current load flow method [1, 6, 16], and the probabilistic load flow method [26–33]. Some of them share certain roots and most of them were dealt with in several algorithms.

In general, the methods for power flow analysis can be divided to deterministic and probabilistic methods. The deterministic methods, such as Newton–Raphson method, Gauss–Seidel method, fast decoupled load flow method, and direct current load flow method use specific values of power generations and load demands of a selected network configuration to calculate system states and power flows. The probabilistic methods require inputs with probability density function to obtain system states and power flows in terms of probability density function, so that the system uncertainties can be included and reflected in the results.

In general, the methods can be divided to direct current methods and alternating current methods. The direct current methods deal only with active power and consider certain additional simplifications, if they are compared with alternating current methods, and are linear methods. The alternating current methods deal with active and reactive power. They are nonlinear methods.

The main objectives of the methods for power flow analysis include the applicability of the methods to large and complex real power systems and the convergence of the iterations [33–36].

The principal information obtained from applying the methods is the magnitude and phase angle of the voltage at each bus and the real and reactive power flowing in each line.

DOI: 10.1007/978-0-85729-688-7_11,

The methods can use the admittance matrix $[\underline{Y}]$, which includes self-bus and mutual admittances, which composes the bus, or they can use the impedance matrix $[\underline{Z}]$, and which includes driving-point and transfer impedances [1, 2]. Let us consider methods using admittances.

11.2 Basis of Power System Model Using Admittance Matrix

The admittance matrix $[\underline{Y}]$ is defined with the following equation:

$$[\underline{Y}] = \left[A_l^T\right] \times [\underline{Y}_{k-l}] \times [A_l] \tag{11.1}$$

The definition of the admittance matrix $[\underline{Y}]$ is preceded by the definition of matrices $[\underline{Y}_{k-l}]$ and $[A_l]$.

Matrix $[\underline{Y}_{k-l}]$ is called a primitive admittance matrix. It is a square matrix in which the number of its rows (columns) is equal to the number of branches of the corresponding graph. Figure 11.1 presents the graph. The subscript used in the nomenclature of the elements of this matrix refers to the end nodes of the corresponding branch. For example, *k–l* refers to the branch between nodes *k* and *l*. The diagonal elements of this matrix are the self-admittances of the corresponding branches, whereas the nondiagonal elements are the admittances of the mutually coupled branches [1, 2].

Let us consider one simple electrical network, depicted with its corresponding scheme (Fig. 11.1).

Figure 11.1 shows a graph, which is called a directed graph for its corresponding electrical power system because each branch is represented between its end nodes by a directed line segment with an arrow in the direction of the branch current. When a branch connects to a node, the branch and node are said to be incident. A tree of graph is formed by those branches of the graph, which interconnect or span all the nodes of the graph without forming any closed path. In general, there are many possible trees of a network because different combinations of branches can be chosen to span the nodes. Thus, for example, one of those combinations is the tree defined by a, b, and d. In that case, all the other branches (c, e, and f) are called links, and when a link is added to a tree, a closed path or loop is formed.

A graph may be described in terms of a connection or incidence matrix. Of particular interest is the branch-to-node incidence matrix $[A_l]$, which has one row for each branch and one column for each node with an entry a_{ij} in row i and column j according to the following rule:

$$\begin{aligned} a_{ij} &= 0 \quad \text{if branch } i \text{ is not connected to node } j \\ a_{ij} &= 1 \quad \text{if current in branch } i \text{ is directed away from node } j \\ a_{ij} &= -1 \quad \text{if current in branch } i \text{ is directed toward from node } j \end{aligned} \tag{11.2}$$

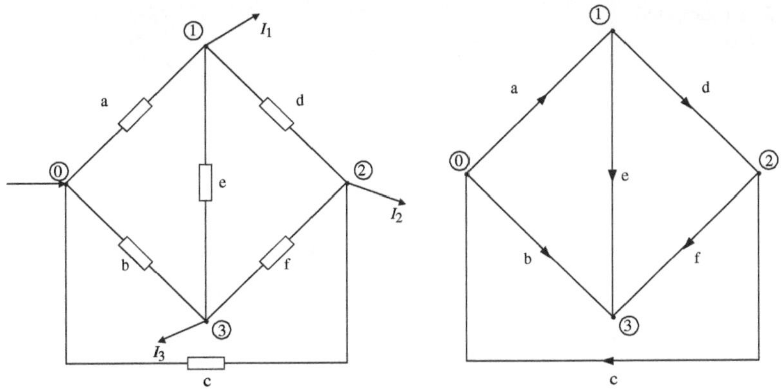

Fig. 11.1 Sample electrical power system (*left*) and its corresponding directed graph (*right*)

In network calculations, we usually choose a slack node or reference node. The column corresponding to the slack node is then omitted from $[A_I]$. For example, choosing node 0 as the slack (Fig. 11.1) and invoking the rule of previous equation, we obtain the rectangular branch-to-node incidence matrix $[A_I]$:

$$[A_I] = \begin{matrix} a \\ b \\ c \\ d \\ e \\ f \end{matrix} \begin{bmatrix} -1 & 0 & 0 \\ 0 & 0 & -1 \\ 0 & 1 & 0 \\ 1 & -1 & 0 \\ 1 & 0 & -1 \\ 0 & 1 & -1 \end{bmatrix}$$
$$\begin{matrix} 1 & 2 & 3 \end{matrix} \quad \rightarrow \text{nodes} \tag{11.3}$$

The non-slack nodes of a network are often called independent nodes or buses.

The second incidence matrix is the branch-to-loop (or branch-to-circuit, or circuit-edge) incidence matrix $[A_{II}]$. An orientation of the directed graph loops should be conducted as the first step of preparing the second incidence matrix. Figure 11.2 shows a directed graph composed of oriented loops. A loop of a directed graph with an orientation assigned by a cyclic ordering of nodes along the circuit is called an oriented loop. The branch orientation corresponds to the direction of the current flowing through the branch, which was already explained while defining the $[A_I]$ matrix.

The orientation of the loops is arbitrary. For example, the loop composed of branches a, b, and e can be oriented as (0, 1, 3) or as (0, 3, 1). Also, we can represent the orientation graphically by an arrowhead. We shall say that the orientations of a branch of a loop and the loop coincide if the branch nodes appear in the same order both in the ordered-pair representation of the branch and in the ordered-node representation of the loop. Otherwise, they are opposite.

Fig. 11.2 An oriented directed graph

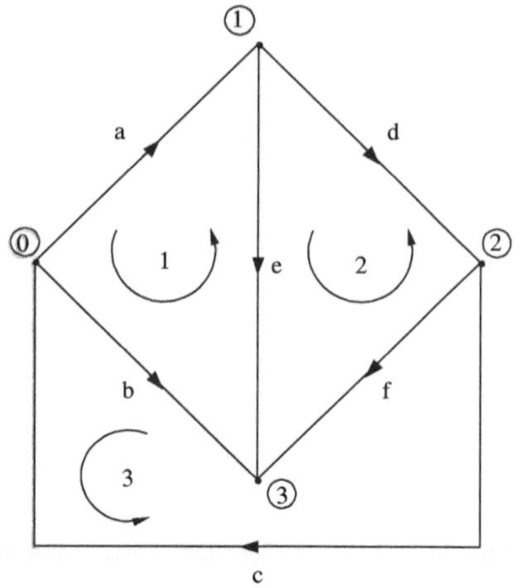

The branch-to-loop incidence matrix $[A_{II}]$ has one row for each branch and one column for each loop with an entry b_{ij} in row i and column j according to the following rule:

$$\begin{aligned} b_{ij} &= 0 \quad \text{if branch } i \text{ is not in loop } j \\ b_{ij} &= 1 \quad \text{if branch } i \text{ is in loop } j \text{ and the orientations} \\ &\qquad \text{of the loop and the branch coincide} \\ b_{ij} &= -1 \quad \text{if branch } i \text{ is in loop } j \text{ and the orientations} \\ &\qquad \text{of the loop and the branch are opposite} \end{aligned} \tag{11.4}$$

Thus, the rectangular branch-to-loop incidence matrix $[A_{II}]$, corresponding to the graph (Fig. 11.2), has the following form:

$$[A_{II}] = \begin{matrix} a \\ b \\ c \\ d \\ e \\ f \end{matrix} \begin{bmatrix} -1 & 0 & 0 \\ 1 & 0 & -1 \\ 0 & 0 & -1 \\ 0 & -1 & 0 \\ -1 & 1 & 0 \\ 0 & -1 & 1 \end{bmatrix} \atop \begin{matrix} 1 & 2 & 3 \end{matrix} \quad \rightarrow \text{loops} \tag{11.5}$$

After defining the admittance matrix $[\underline{Y}]$, the starting point of the analysis is the single line diagram of the system. Power lines and transformers are represented by their per-phase nominal—π equivalent circuits [2]. Shunt conductance G is usually neglected in overhead power lines when calculating voltage and current.

For each line and transformer, numerical values for the series impedance $\underline{Z}$ and the total admittance $\underline{Y}$ are necessary to determine all the elements of the $N \times N$ admittance matrix of which the typical element $\underline{Y}_{ij}$ is:

$$\underline{Y}_{ij} = Y_{ij} \cos \delta_{ij} + jY_{ij} \sin \delta_{ij} = G_{ij} + jB_{ij}, \tag{11.6}$$

where $\underline{Y}_{ij}$ is the element $i - j$ of the admittance matrix, Y_{ij} is the magnitude of the admittance matrix $i - j$ element, δ_{ij} is the phase angle of the admittance matrix $i - j$ element, G_{ij} is the conductance of the admittance matrix $i - j$ element, and B_{ij} is the susceptance of the admittance matrix $i - j$ element.

The voltage at a typical bus i of the system is given in polar coordinates by:

$$\underline{U}_i = U_i \cos \theta_i + jU_i \sin \theta_i, \tag{11.7}$$

where $\underline{U}_i$ is the complex voltage in bus i, U_i is the magnitude of the voltage in bus i, and θ_i is the phase angle of the voltage in bus i.

The voltage at another bus j is similarly written by changing the subscript from i to j. The net current injected into the network at bus i in terms of the elements $\underline{Y}_{ij}$ of admittance matrix $[\underline{Y}]$ is given by the sum:

$$\underline{I}_i = \underline{Y}_{i1}\underline{U}_1 + \underline{Y}_{i2}\underline{U}_2 + \cdots + \underline{Y}_{iN}\underline{U}_N = \sum_{j=1}^{N} \underline{Y}_{ij}\underline{U}_j \tag{11.8}$$

Let P_i and Q_i denote the net real and reactive power entering the network at the bus i. Then, the complex power injected at bus i is:

$$\underline{S}_i = P_i + jQ_i = U_i \times \underline{I}_i^* = U_i \times \sum_{j=1}^{N} \underline{Y}_{ij}^* \underline{U}_j^* \tag{11.9}$$

Expanding previous equation with preceding two equations and presenting it in the exponential form rather than in the trigonometric form of the polar coordinates, we obtain:

$$\begin{aligned} P_i + jQ_i &= U_i \times e^{j\theta_i} \times \sum_{j=1}^{N} Y_{ij} \times e^{-j\delta_{ij}} \times U_j \times e^{-j\theta_j} \\ &= U_i \times \sum_{j=1}^{N} Y_{ij} \times U_j \times e^{-j(\delta_{ij}+\theta_j-\theta_i)} \end{aligned} \tag{11.10}$$

Equating the real and reactive parts, we obtain:

$$P_i = U_i \times \sum_{j=1}^{N} Y_{ij} \times U_j \times \cos(\delta_{ij} + \theta_j - \theta_i) \tag{11.11}$$

$$Q_i = -U_i \times \sum_{j=1}^{N} Y_{ij} \times U_j \times \sin(\delta_{ij} + \theta_j - \theta_i) \tag{11.12}$$

Fig. 11.3 Active and reactive power notation at typical bus i

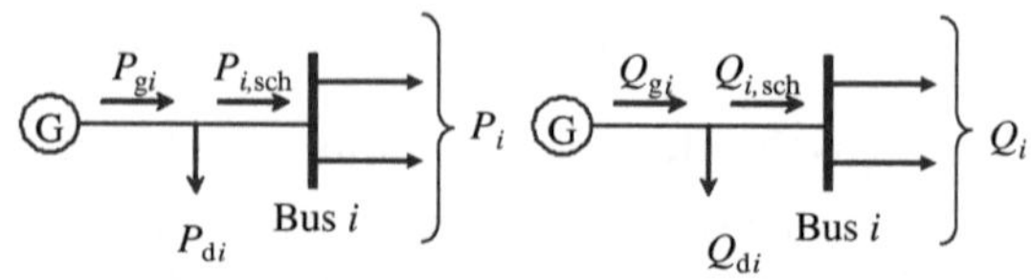

This equation constitutes the polar form of the power-flow equations. They provide calculated values for the net real power P_i and reactive power Q_i entering the network at typical bus i. Let P_{gi} denote the scheduled power being generated at bus i and P_{di} denote the scheduled power demand of the load at that bus (Fig. 11.3). Then, $P_{i,\text{sch}} = P_{gi} - P_{di}$ is the net scheduled power being injected into the network at bus i.

Denoting the calculated value of P_i by $P_{i,\text{calc}}$ leads to the definition of mismatch ΔP_i:

$$\Delta P_i^{(0)} = P_{i,\text{sch}} - P_{i,\text{calc}}^{(0)} = (P_{gi} - P_{di}) - P_{i,\text{calc}} \tag{11.13}$$

Likewise, for reactive power at bus i, we have:

$$\Delta Q_i^{(0)} = Q_{i,\text{sch}} - Q_{i,\text{calc}}^{(0)} = (Q_{gi} - Q_{di}) - Q_{i,\text{calc}} \tag{11.14}$$

Mismatches occur in the course of solving a power-flow problem when calculated values of P_i and Q_i do not coincide with the scheduled values.

It can be easily concluded that, in general, four potentially unknown quantities are associated with each bus i: P_i, Q_i, U_i, and θ_i. The general practice in power-flow studies is to identify three types of buses in the network. At each bus i, two of the four above-mentioned quantities are specified and the remaining two are calculated. Specified quantities are chosen according to the following categorization of buses:

- *Load buses*: At each nongenerator bus, which is called a load bus, both P_{gi} and Q_{gi} are zero and the real power P_{di} and reactive power Q_{di} drawn from the system by the load are known from historical record, load forecast, or measurement. A load bus i is often called a $P - Q$ bus because the scheduled values $P_{i,\text{sch}} = -P_{di}$ and $Q_{i,\text{sch}} = -Q_{di}$ are known and mismatches ΔP_i and ΔQ_i can be defined. Thus, the remaining two unknown quantities to be determined for the bus are U_i and θ_i.
- *Voltage-controlled buses*: Any bus of the system at which the voltage magnitude is kept constant is said to be voltage controlled. At each bus, to which there is a generator connected, the generation can be controlled by adjusting the prime mover, and the voltage magnitude can be controlled by adjusting the generation excitation. Therefore, at each generator bus i, we may properly specify P_{gi} and U_i. With P_{di} also known, we can define mismatch ΔP_i according to Eq. 11.13. Generator reactor power Q_{gi} required to support the scheduled voltage U_i cannot be known in advance and so mismatch ΔQ_i is not defined. Therefore, at a

generator bus i, θ_i and Q_i are the unknown quantities. For obvious reasons, a generator bus is usually called a $P - V$ bus. Certain buses without generators may also have voltage control capability; such buses are also designated voltage-controlled buses at which the real power generation is zero.

- *Slack bus*: This is the bus to which the regulating power plant (generator) of the system is connected. The voltage angle of the slack bus serves as a reference for the angles of all other bus voltages. Thus, U_i and θ_i are specified for the slack bus.

The functions P_i and Q_i of Eqs. 11.11 and 11.12, respectively, are nonlinear functions of state variables U_i and θ_i. Hence, power-flow calculations usually employ iterative techniques, such as the Gauss-Seidel and Newton–Raphson procedures. The Newton–Raphson method solves the polar form of the power-flow equations until the ΔP_i and ΔQ_i mismatches at all buses fall within specified tolerances. The Gauss–Seidel method solves the power flow equations in rectangular (complex variable) coordinates until differences in bus voltages from one iteration to another are sufficiently small. Both methods are based on bus admittance equations.

11.3 Newton–Raphson Method

Taylor series expansion for a function of two or more variables is the basis for the Newton–Raphson method of solving the power flow problem [1–4]. To apply the Newton–Raphson method to the solution of the power-flow equations, we express bus voltages and line admittances in polar form. When j is set to i in Eq. 11.11 and the corresponding terms are separated from the sum, we obtain:

$$P_{i,\text{calc}} = P_i = G_{ii} \times U_i^2 + U_i \times \sum_{\substack{j=1 \\ j \neq i}}^{N} Y_{ij} \times U_j \times \cos(\delta_{ij} + \theta_j - \theta_i) \tag{11.15}$$

$$Q_{i,\text{calc}} = Q_i = -B_{ii} \times U_i^2 - U_i \times \sum_{\substack{j=1 \\ j \neq i}}^{N} Y_{ij} \times U_j \times \cos(\delta_{ij} + \theta_j - \theta_i) \tag{11.16}$$

Considering Eq. 11.6 and applying $\theta_i - \theta_j = \theta_{ij}$, we obtain:

$$P_i = G_{ii} \times U_i^2 + U_i \times \sum_{\substack{j=1 \\ j \neq i}}^{N} U_j \times \left(G_{ij} \cos\theta_{ij} + B_{ij} \sin\theta_{ij}\right) \tag{11.17}$$

$$Q_i = -B_{ii} \times U_i^2 + U_i \times \sum_{\substack{j=1 \\ j \neq i}}^{N} U_j \times \left(G_{ij} \sin\theta_{ij} - B_{ij} \cos\theta_{ij}\right) \tag{11.18}$$

These Eqs. 11.17 and 11.18 can be readily differentiated with respect to voltage angles and magnitudes.

Let us consider a power network with N buses. For one of them, the slack bus, U_i and θ_i are specified. Let us denote with q the number of buses for which U_i and θ_i are unknown quantities. Thus, in solving the power-flow problem, i.e., calculating the complex bus voltages, one should calculate the following unknown quantities:

- $N - 1$ unknown voltage angles θ_i
- q unknown voltage magnitudes

Thus, the total number of unknown variables is $N - 1 + q$ and these variables are referred to as state variables because their values, which describe the state of the system, depend on the quantities specified at all the buses. Hence, the power flow problem is to determine values for all state variables by solving an equal number of power-flow equations based on the input data specifications. They can be determined by solving a system of $N - 1 + q$ equations, $N - 1$ of which are of type like Eq. 11.15 and the other q are of type like Eq. 11.16. Instead of equations of type like Eq. 11.15, one can use equations of type like Eq. 11.17. Similarly, instead using equations of type like Eq. 11.16, one can use equations of type like Eq. 11.18. The system of equations formed in this way is a nonlinear system of equations.

The slack bus is assumed as the Nth bus. The subscripts from 1 to q are correlated to the load buses.

Considering the Eqs. 11.15 and 11.16, i.e., 11.17 and 11.18, it is obvious that they are functions of the voltage angles $\theta_1, \theta_2, \ldots, \theta_{N-1}$ as well as of voltage magnitudes $U_1, U_2, \ldots, U_q$. Therefore, formally it can be written:

$$P_i = P_i(\theta_1, \theta_2, \ldots, \theta_{N-1}, U_1, U_2, \ldots, U_q) \tag{11.19}$$

$$Q_i = Q_i(\theta_1, \theta_2, \ldots, \theta_{N-1}, U_1, U_2, \ldots, U_q) \tag{11.20}$$

Before the start of the iterative calculations, we assign initial values to all unknown variables: $\theta_1^{(0)}, \theta_2^{(0)}, \ldots, \theta_{N-1}^{(0)}, U_1^{(0)}, U_2^{(0)}, \ldots, U_q^{(0)}$. These are assumed initial values and are differing from the exact ones. The mismatches between the exact and the assumed values are denoted with $\Delta\theta_1^{(0)}, \Delta\theta_2^{(0)}, \ldots, \Delta\theta_{N-1}^{(0)}, \Delta U_1^{(0)}, \Delta U_2^{(0)}, \ldots, \Delta U_q^{(0)}$, respectively. It is recommended that the initial values of all the voltage angles should be set equal to zero and all the voltage magnitudes should be set equal to 1 [p. u.] or to be set equal to the voltage magnitude of the slack bus.

Equations 11.19 and 11.20 can be rewritten:

$$P_i = P_i\left(\theta_1^{(0)} + \Delta\theta_1^{(0)}, \ldots, \theta_{N-1}^{(0)} + \Delta\theta_{N-1}^{(0)}, U_1^{(0)} + \Delta U_1^{(0)}, \ldots, U_q^{(0)} + \Delta U_q^{(0)}\right) \tag{11.21}$$

$$Q_i = Q_i\left(\theta_1^{(0)} + \Delta\theta_1^{(0)}, \ldots, \theta_{N-1}^{(0)} + \Delta\theta_{N-1}^{(0)}, U_1^{(0)} + \Delta U_1^{(0)}, \ldots, U_q^{(0)} + \Delta U_q^{(0)}\right) \tag{11.22}$$

Expanding the Eq. 11.21 in Taylor series, the following equation is obtained:

$$\begin{aligned} P_i = {} & P_i\left(\theta_1^{(0)}, \theta_2^{(0)}, \ldots, \theta_{N-1}^{(0)}, U_1^{(0)}, U_2^{(0)}, \ldots, U_q^{(0)}\right) \\ & + \frac{\partial P_i}{\partial \theta_1}\bigg|^{(0)} \times \Delta\theta_1^{(0)} + \frac{\partial P_i}{\partial \theta_2}\bigg|^{(0)} \times \Delta\theta_2^{(0)} + \cdots + \frac{\partial P_i}{\partial \theta_{N-1}}\bigg|^{(0)} \times \Delta\theta_{N-1}^{(0)} \\ & + \frac{\partial P_i}{\partial U_1}\bigg|^{(0)} \times \Delta U_1^{(0)} + \frac{\partial P_i}{\partial U_2}\bigg|^{(0)} \times \Delta U_2^{(0)} + \cdots + \frac{\partial P_i}{\partial U_q}\bigg|^{(0)} \times \Delta U_q^{(0)} + \mathrm{Re} \end{aligned} \tag{11.23}$$

In Eq. 11.23, the term $(\partial P_i/\partial \theta_i)^{(0)}$ indicates that the corresponding partial derivative is evaluated for the assumed value of voltage angle θ_1 in the 0th iteration, $\theta_1^{(0)}$. Other such terms are evaluated similarly. The last term, Re, stands for the remainder of the Taylor series expansion, which comprises partial derivatives of second-order and higher.

The term Re can be neglected, if the mismatches between the exact and the assumed values are small. Considering Eqs. 11.13 and 11.23, the following equation is obtained:

$$\begin{aligned} \Delta P_i^{(0)} = {} & \frac{\partial P_i}{\partial \theta_1}\bigg|^{(0)} \times \Delta\theta_1^{(0)} + \frac{\partial P_i}{\partial \theta_2}\bigg|^{(0)} \times \Delta\theta_2^{(0)} + \cdots + \frac{\partial P_i}{\partial \theta_{N-1}}\bigg|^{(0)} \times \Delta\theta_{N-1}^{(0)} \\ & + \frac{\partial P_i}{\partial U_1}\bigg|^{(0)} \times \Delta U_1^{(0)} + \frac{\partial P_i}{\partial U_2}\bigg|^{(0)} \times \Delta U_2^{(0)} + \cdots + \frac{\partial P_i}{\partial U_q}\bigg|^{(0)} \times \Delta U_q^{(0)} \end{aligned} \tag{11.24}$$

Analogically, by expanding the Eq. 11.22 in Taylor series, the following equation is obtained:

$$\begin{aligned} Q_i = {} & Q_i\left(\theta_1^{(0)}, \theta_2^{(0)}, \ldots, \theta_{N-1}^{(0)}, U_1^{(0)}, U_2^{(0)}, \ldots, U_q^{(0)}\right) \\ & + \frac{\partial Q_i}{\partial \theta_1}\bigg|^{(0)} \times \Delta\theta_1^{(0)} + \frac{\partial Q_i}{\partial \theta_2}\bigg|^{(0)} \times \Delta\theta_2^{(0)} + \cdots + \frac{\partial Q_i}{\partial \theta_{N-1}}\bigg|^{(0)} \times \Delta\theta_{N-1}^{(0)} \\ & + \frac{\partial Q_i}{\partial U_1}\bigg|^{(0)} \times \Delta U_1^{(0)} + \frac{\partial Q_i}{\partial U_2}\bigg|^{(0)} \times \Delta U_2^{(0)} + \cdots + \frac{\partial Q_i}{\partial U_q}\bigg|^{(0)} \times \Delta U_q^{(0)} + \mathrm{Re} \end{aligned} \tag{11.25}$$

If the mismatches between the exact and the assumed values are small and considering Eqs. 11.14 and 11.25, the following equation is obtained:

$$\Delta Q_i^{(0)} = \frac{\partial Q_i}{\partial \theta_1}\bigg|^{(0)} \times \Delta\theta_1^{(0)} + \frac{\partial Q_i}{\partial \theta_2}\bigg|^{(0)} \times \Delta\theta_2^{(0)} + \cdots + \frac{\partial Q_i}{\partial \theta_{N-1}}\bigg|^{(0)} \times \Delta\theta_{N-1}^{(0)}$$
$$+ \frac{\partial Q_i}{\partial U_1}\bigg|^{(0)} \times \Delta U_1^{(0)} + \frac{\partial Q_i}{\partial U_2}\bigg|^{(0)} \times \Delta U_2^{(0)} + \cdots + \frac{\partial Q_i}{\partial U_q}\bigg|^{(0)} \times \Delta U_q^{(0)} \quad (11.26)$$

Equation 11.24 can be written for each of the $N - 1$ buses, for which $P_{i,\mathrm{sch}}$ is known. Similarly, Eq. 11.26 can be written for each of the q buses, for which $Q_{i,\mathrm{sch}}$ is known. In that way, a system of $N - 1 + q$ independent equations is formed. The number of equations is equal to the number of unknown variables. By solving the system of equations, one can obtain the corrections $\Delta\theta_1^{(0)}$, $\Delta\theta_2^{(0)}$, . . . , $\Delta\,\theta_{N-1}^{(0)}$, $\Delta U_1^{(0)}$, $\Delta\, U_2^{(0)}$, . . . , $\Delta U_q^{(0)}$. Thus, the values of the unknown quantities after the first iteration are:

$$\begin{aligned}
\theta_1^{(1)} &= \theta_1^{(0)} + \Delta\theta_1^{(0)}\\
\theta_2^{(1)} &= \theta_2^{(0)} + \Delta\theta_2^{(0)}\\
&\vdots\\
\theta_{N-1}^{(1)} &= \theta_{N-1}^{(0)} + \Delta\theta_{N-1}^{(0)}
\end{aligned} \quad (11.27)$$

$$\begin{aligned}
U_1^{(1)} &= U_1^{(0)} + \Delta U_1^{(0)}\\
U_2^{(1)} &= U_2^{(0)} + \Delta U_2^{(0)}\\
&\vdots\\
U_q^{(1)} &= U_q^{(0)} + \Delta U_q^{(0)}
\end{aligned} \quad (11.28)$$

By using these new (actual) values of the unknown variables, the new values for P_i and Q_i are calculated by using the Eqs. 11.15 or 11.17 and 11.16 or 11.18, respectively. The new mismatches $\Delta P_i^{(1)}$ for $i = 1, \ldots, N - 1$ and $\Delta Q_i^{(1)}$ *for* $i = 1, \ldots, q$ are calculated using these new values for P_i and Q_i. The superscript $^{(1)}$ stands for the first iteration.

After each iteration, it is checked if the condition for termination of the iterative calculation is satisfied. For this reason, the absolute values of the mismatches $\Delta P_i^{(v)}$ $i = 1, \ldots, N - 1$ and $\Delta Q_i^{(v)}$ for $i = 1, \ldots, q$ are calculated in each iteration $v = 1, 2, \ldots$ When these absolute values are smaller than a predefined tolerance ε, the iterative process is stopped, e.g., in the vth iteration:

$$\max_{i=1,\ldots,N-1}\left\{\left|\Delta P_i^{(v)}\right|\right\} \leq \varepsilon \quad (11.29)$$

$$\max_{i=1,\ldots,q}\left\{\left|\Delta Q_i^{(v)}\right|\right\} \leq \varepsilon \quad (11.30)$$

The iterative process can be stopped and the values of $\theta_1^{(v)}, \theta_2^{(v)}, \ldots, \theta_{N-1}^{(v)}, U_1^{(v)}, U_2^{(v)}, \ldots, U_q^{(v)}$ are accounted as accurate enough to represent the searched solution. The value of ε is determined in correspondence with the accuracy of the input data and the intended application of the results. Thus, the iterative process continues until the relations in Eqs. 11.29 and 11.30 are satisfied.

The system of $N - 1 + q$ equations that is calculated in each of the iterations can be presented in matrix form:

$$\begin{bmatrix} \frac{\partial P_1}{\partial \theta_1} & \cdots & \frac{\partial P_1}{\partial \theta_{N-1}} & \frac{\partial P_1}{\partial U_1} & \cdots & \frac{\partial P_1}{\partial U_q} \\ \vdots & \cdots & \vdots & \vdots & \cdots & \vdots \\ \frac{\partial P_{N-1}}{\partial \theta_1} & \cdots & \frac{\partial P_{N-1}}{\partial \theta_{N-1}} & \frac{\partial P_{N-1}}{\partial U_1} & \cdots & \frac{\partial P_{N-1}}{\partial U_q} \\ \frac{\partial Q_1}{\partial \theta_1} & \cdots & \frac{\partial Q_1}{\partial \theta_{N-1}} & \frac{\partial Q_1}{\partial U_1} & \cdots & \frac{\partial Q_1}{\partial U_q} \\ \vdots & \cdots & \vdots & \vdots & \cdots & \vdots \\ \frac{\partial Q_q}{\partial \theta_1} & \cdots & \frac{\partial Q_q}{\partial \theta_{N-1}} & \frac{\partial Q_q}{\partial U_1} & \cdots & \frac{\partial Q_q}{\partial U_q} \end{bmatrix} \times \begin{bmatrix} \Delta\theta_1 \\ \vdots \\ \Delta\theta_{N-1} \\ \Delta U_1 \\ \vdots \\ \Delta U_q \end{bmatrix} = \begin{bmatrix} \Delta P_1 \\ \vdots \\ \Delta P_{N-1} \\ \Delta Q_1 \\ \vdots \\ \Delta Q_q \end{bmatrix} \tag{11.31}$$

The matrix Eq. 11.31 can be written in more suitable form:

$$\underbrace{\begin{bmatrix} \frac{\partial P_1}{\partial \theta_1} & \cdots & \frac{\partial P_1}{\partial \theta_{N-1}} & \frac{\partial P_1}{\partial U_1}U_1 & \cdots & \frac{\partial P_1}{\partial U_q}U_q \\ \vdots & H_{ij} & \vdots & \vdots & N_{ij} & \vdots \\ \frac{\partial P_{N-1}}{\partial \theta_1} & \cdots & \frac{\partial P_{N-1}}{\partial \theta_{N-1}} & \frac{\partial P_{N-1}}{\partial U_1}U_1 & \cdots & \frac{\partial P_{N-1}}{\partial U_q}U_q \\ \frac{\partial Q_1}{\partial \theta_1} & \cdots & \frac{\partial Q_1}{\partial \theta_{N-1}} & \frac{\partial Q_1}{\partial U_1}U_1 & \cdots & \frac{\partial Q_1}{\partial U_q}U_q \\ \vdots & M_{ij} & \vdots & \vdots & L_{ij} & \vdots \\ \frac{\partial Q_q}{\partial \theta_1} & \cdots & \frac{\partial Q_q}{\partial \theta_{N-1}} & \frac{\partial Q_q}{\partial U_1}U_1 & \cdots & \frac{\partial Q_q}{\partial U_q}U_q \end{bmatrix}}_{\text{Jacobian}} \times \underbrace{\begin{bmatrix} \Delta\theta_1 \\ \vdots \\ \Delta\theta_{N-1} \\ \frac{\Delta U_1}{U_1} \\ \vdots \\ \frac{\Delta U_q}{U_q} \end{bmatrix}}_{\text{Corrections}} = \underbrace{\begin{bmatrix} \Delta P_1 \\ \vdots \\ \Delta P_{N-1} \\ \Delta Q_1 \\ \vdots \\ \Delta Q_q \end{bmatrix}}_{\text{Mismatches}} \tag{11.32}$$

At this point, it is usual praxis that the Jacobian matrix is divided into four sub-matrices, denoted with H, N, M, and L. The elements of these sub-matrices are defined as follows:

$$H_{ij} = \frac{\partial P_i}{\partial \theta_j}; \quad i = 1,\ldots,N-1; \quad j = 1,\ldots,N-1 \tag{11.33}$$

$$N_{ij} = \frac{\partial P_i}{\partial U_j} U_j; \quad i = 1,\ldots,N-1; \quad j = 1,\ldots,q \tag{11.34}$$

$$M_{ij} = \frac{\partial Q_i}{\partial \theta_j}; \quad i = 1,\ldots,q; \quad j = 1,\ldots,N-1 \tag{11.35}$$

$$L_{ij} = \frac{\partial Q_i}{\partial U_j} U_j; \quad i = 1,\ldots,q; \quad j = 1,\ldots,q \tag{11.36}$$

Additionally, if following expressions are established:

$$\Delta P = \begin{bmatrix} \Delta P_1 \\ \Delta P_2 \\ \vdots \\ \Delta P_{N-1} \end{bmatrix}, \quad \Delta Q = \begin{bmatrix} \Delta Q_1 \\ \Delta Q_2 \\ \vdots \\ \Delta Q_q \end{bmatrix}, \quad \Delta\theta = \begin{bmatrix} \Delta\theta_1 \\ \Delta\theta_2 \\ \vdots \\ \Delta\theta_{N-1} \end{bmatrix} \quad \Delta U/U = \begin{bmatrix} \Delta U_1/U_1 \\ \Delta U_2/U_2 \\ \vdots \\ \Delta U_q/U_q \end{bmatrix} \tag{11.37}$$

The matrix Eq. 11.32 can be written in the following form considering the expression 11.37:

$$\begin{bmatrix} H & N \\ M & L \end{bmatrix} \times \begin{bmatrix} \Delta\theta \\ \Delta U/U \end{bmatrix} = \begin{bmatrix} \Delta P \\ \Delta Q \end{bmatrix} \tag{11.38}$$

The calculation process in each iteration of the Newton–Raphson power flow solution is presented with the matrix Eq. 11.38. The values of the unknown variables in the νth iteration are calculated as follows:

$$\begin{aligned}
\theta_1^{(\nu)} &= \theta_1^{(\nu-1)} + \Delta\theta_1^{(\nu-1)} \\
\theta_2^{(\nu)} &= \theta_2^{(\nu-1)} + \Delta\theta_2^{(\nu-1)} \\
&\vdots \\
\theta_{N-1}^{(\nu)} &= \theta_{N-1}^{(\nu-1)} + \Delta\theta_{N-1}^{(\nu-1)} \\
U_1^{(\nu)} &= U_1^{(\nu-1)} + \Delta U_1^{(\nu-1)} \\
U_2^{(\nu)} &= U_2^{(\nu-1)} + \Delta U_2^{(\nu-1)} \\
&\vdots \\
U_q^{(\nu)} &= U_q^{(\nu-1)} + \Delta U_q^{(\nu-1)}
\end{aligned} \tag{11.39}$$

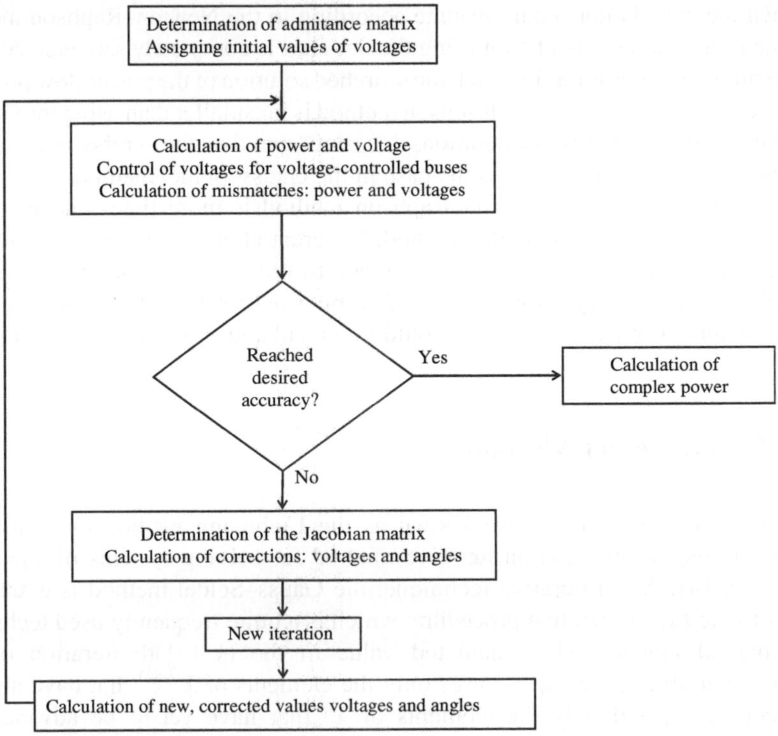

Fig. 11.4 Newton–Raphson method

The iterative process is stopped when the relations (11.29) and (11.30) are satisfied.

Figure 11.4 shows the calculation algorithm associated to the Newton–Raphson method in solving a power flow problem.

11.3.1 General Characteristics of the Newton–Raphson Method

Few points should be emphasized considering the main characteristics of the Newton–Raphson method for calculating power flow. The most important is the fact that the efficiency of this method depends on the initial values of the unknown variables, i.e., it is efficient if these initial values that we assign to the unknown variables at the beginning of the iterative process do not differ from their actual values in a large extent. If the initial values are selected well close to the actual values, the method is efficient. Some authors suggest that the initial two to three iterations should be calculated using the Gauss–Seidel method after which the iterative process should resume using the Newton–Raphson method [3]. Or some other authors suggest that the initial values should be determined using an approximate noniterative estimation and then the calculation should resume according to the Newton–Raphson method [4]. The initial values calculation using the flat-start is completely satisfactory.

The number of iterations in which the searched solution of the power flow problem is reached by using the Newton–Raphson method is far smaller than using the Gauss–Seidel method. The number of iterations does not depend on the number of buses and the topology of the network, as is the case in the Gauss–Seidel method.

On the other side, the Newton–Raphson method is more time- and memory-consuming than the Gauss–Seidel method. Different attempts and modifications of the Newton–Raphson method are being made to reduce the calculation time. For example, there are suggestions that the Jacobian matrix from (11.32), instead of being calculated in each iteration, should be calculated in each second iteration.

11.4 Gauss–Seidel Method

The Gauss–Seidel method, also known as the Liebmann method or method of successive displacement, is an iterative method for solving systems of equations [1, 2, 5, 9, 14]. As an iterative technique, the Gauss–Seidel method is extremely similar to the Jacobi iterative procedure, which is a quite frequently used technique in numerical algebra. The calculated value in the $(k + 1)$th iteration of the unknown variable x_i, i.e., $x_i^{(k+1)}$, uses only the elements of $X^{(k+1)}$ that have already been computed, and only the elements of X^k that have yet to be advanced to iteration $k + 1$. This means that unlike the Jacobi method, only one storage vector is required as elements can be overwritten as they are computed, which can be advantageous for very large problems.

Let us consider a power network with N buses. The slack bus is denoted with index s. Considering Eq. 11.9, the complex value of the current injected into the network at bus i is:

$$\underline{I}_i^* = \left(\frac{\underline{S}_i}{\underline{U}_i}\right)^* = \frac{P_i - jQ_i}{\underline{U}_i^*} \tag{11.40}$$

Then, considering Eq. 11.8 and assuming flat start conditions regarding the initial values of the unknown bus voltages, the following expression can be derived and used in calculating the complex voltage in bus i in the $(v + 1)$ iteration:

$$\underline{U}_i^{(v+1)} = \frac{1}{\underline{Y}_{ii}}\left(\frac{P_i - jQ_i}{(\underline{U}_i^{(v)})^*} - \sum_{l=1}^{i-1} \underline{Y}_{il}\underline{U}_l^{(v+1)} - \sum_{l=i+1}^{N} \underline{Y}_{il}\underline{U}_l^{(v)}\right), \quad i = 1, \ldots, n; \quad i \neq s \tag{11.41}$$

A control check whether the iterative procedure can be terminated is conducted at the end of each iteration. For that purpose, one of the options is to check the complex voltage correction, $\Delta \underline{U}_i^{(v+1)}$, with the highest module:

$$\max_i \left\{ \left| \Delta \underline{U}_i^{(v+1)} \right| \right\} = \max_i \left\{ \left| \Delta \underline{U}_i^{(v+1)} - \Delta \underline{U}_i^{(v)} \right| \right\} \leq \varepsilon \tag{11.42}$$

If this module is less than the specified tolerance ε, that is predefined prior starting the iterative procedure as in the case of the Newton–Raphson method, the iterative procedure can be stopped and the calculated values of the bus voltages $\underline{U}_i$ in the last ongoing iteration are accounted as accurate enough to represent the searched solution. Instead of the previous criterion for termination of the iterative procedure, one can also use a slightly modified criterion. According to this, latter criterion, first the absolute values of the real and imaginary parts of the complex bus voltages are determined. Then, a control check is conducted whether the maximum of these absolute values is less than the tolerance ε:

$$\max_i \left\{ \left| \mathrm{Re}\left(\Delta \underline{U}_i^{(v+1)} \right) \right| \right\} = \max_i \left\{ \left| \mathrm{Re}\left(\Delta \underline{U}_i^{(v+1)} - \Delta \underline{U}_i^{(v)} \right) \right| \right\} \leq \varepsilon \tag{11.43}$$

$$\max_i \left\{ \left| \mathrm{Im}\left(\Delta \underline{U}_i^{(v+1)} \right) \right| \right\} = \max_i \left\{ \left| \mathrm{Im}\left(\Delta \underline{U}_i^{(v+1)} - \Delta \underline{U}_i^{(v)} \right) \right| \right\} \leq \varepsilon \tag{11.44}$$

If these conditions in Eqs. 11.43 and 11.44 are fulfilled, then the iterative procedure can be stopped and the calculated values of the bus voltages U_i in the last iteration are accounted as the searched solution. It should be noted that for same ε, conditions in Eq. 11.42 and both Eqs. 11.43 and 11.44 are not equal. The former one is more stringent.

If there are voltage-controlled buses along with the slack bus and the load buses, some supplementary relations associated only to the voltage-controlled buses should be incorporated so that the Gauss–Seidel method can be applicable.

For Eq. 11.41 to be applicable for the voltage-controlled buses, the injected reactive power at those buses should be determined. Therefore, the actual complex bus voltages as well as the admittance matrix $[\underline{Y}]$ are used in deriving the mentioned reactive power. Let us assume that bus i is one of the voltage-controlled buses for which the voltage magnitude should be U_i. Considering Eqs. 11.9 and 11.16, the reactive power θ_i entering the network at this voltage-controlled bus i in the $(v + 1)$th iteration can be calculated given one of the following relations:

$$Q_i^{(v+1)} = \mathrm{Im}\left\{ \underline{U}_i^{(v)} \times \sum_{l=1}^{i-1} \left(\underline{Y}_{il} \underline{U}_l^{(v+1)} \right)^* + \underline{U}_i^{(v)} \times \sum_{l=i}^{N} \left(\underline{Y}_{il} \underline{U}_l^{(v+1)} \right)^* \right\} \tag{11.45}$$

or

$$Q_i^{(v+1)} = U_i^{(v)} \times \sum_{l=1}^{i-1} U_l^{(v+1)} \times \left[G_{il} \times \sin\left(\theta_i^{(v)} - \theta_l^{(v+1)}\right) - B_{il} \times \cos\left(\theta_i^{(v)} - \theta_l^{(v+1)}\right)\right]$$
$$+ U_i^{(v)} \times \sum_{l=i}^{N} U_l^{(v)} \times \left[G_{il} \times \sin\left(\theta_i^{(v)} - \theta_l^{(v)}\right) - B_{il} \times \cos\left(\theta_i^{(v)} - \theta_l^{(v)}\right)\right] \tag{11.46}$$

In Eq. 11.46, θ_i stands for the phase angle of the complex voltage at bus i. The assumed value of the phase angle is assigned to θ_i in the first iteration.

Then, a control check is conducted whether the injected reactive power at bus i is within definite limits given by the inequality:

$$Q_{i_{\min}} \leq Q_i^{(v+1)} \leq Q_{i_{\max}} \tag{11.47}$$

where $Q_{i_{\min}}$ is the minimum and $Q_{i_{\max}}$ is the maximum limit imposed on the reactive power output of the generator at the specified voltage-controlled bus i.

In the course of power-flow calculations, if the calculated value of $Q_i^{(v+1)}$ is outside either limit, then $Q_i^{(v+1)}$ is set equal to the limit violated. Namely, if it is determined in the course of calculating the $Q_i^{(v+1)}$ that

$$Q_i^{(v+1)} \leq Q_{i_{\min}} \tag{11.48}$$

then in the subsequent calculations of the actual ongoing iteration, it is set:

$$Q_i^{(v+1)} = Q_{i_{\min}} \tag{11.49}$$

Analogically, if it is determined that

$$Q_i^{(v+1)} \geq Q_{i_{\max}} \tag{11.50}$$

then in the subsequent calculations of the actual ongoing iteration, it is set:

$$Q_i^{(v+1)} = Q_{i_{\max}} \tag{11.51}$$

In the subsequent calculations in iteration $(v + 1)$, the calculated value of the reactive power Q_i entering the network at the voltage-controlled bus i is used for determination of the phase angle of the voltage at the same bus, θ_i. For that purpose, on the basis of Eq. 11.41, the real and the imaginary part of the complex voltage for the specified bus are being calculated first, i.e.:

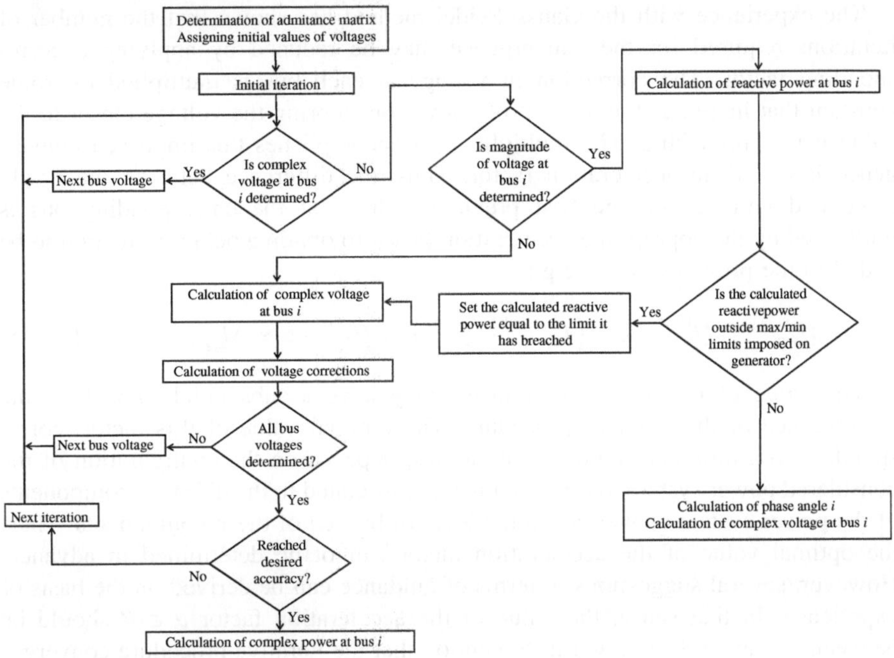

Fig. 11.5 Gauss–Seidel method

$$\underline{U}_i^{(v+1)} = E_i^{(v+1)} + jF_i^{(v+1)} = \frac{1}{\underline{Y}_{ii}} \left(\frac{P_i - jQ_i}{\left(\underline{U}_i^{(v)}\right)^*} - \sum_{l=1}^{i-1} \underline{Y}_{il} \underline{U}_l^{(v+1)} - \sum_{l=i+1}^{N} \underline{Y}_{il} \underline{U}_l^{(v)} \right) \tag{11.52}$$

Subsequently, the phase angle is determined using the following equation:

$$\theta_i^{(v+1)} = \arctan \left(\frac{F_i^{(v+1)}}{E_i^{(v+1)}} \right) \tag{11.53}$$

after which the complex voltage for the specified voltage-controlled bus in iteration $(v + 1)$ is finally being derived as:

$$\underline{U}_i^{(v+1)} = U_i \times \left(\sin \theta_i^{(v+1)} + j \cos \theta_i^{(v+1)} \right) \tag{11.54}$$

Figure 11.5 shows the calculation algorithm associated to the Gauss–Seidel method in solving a power flow problem.

The experience with the Gauss–Seidel method has shown that the number of iterations required for the convergence may be reduced by applying a corresponding factor. The correction in voltage at each bus is multiplied by some constant that increases the amount of correction to bring the voltage closer to the value it is approaching. The multiplier that accomplishes this improved convergence is called an acceleration factor. Thus, the difference between the newly calculated voltage and the best previous voltage at the corresponding bus is multiplied by the appropriate acceleration factor to obtain a better correction to be added to the previous value, e.g.:

$$\underline{U}_i^{(\nu+1)} = \underline{U}_i^{(\nu)} + \alpha \times \left(\underline{U}_i^{(\nu+1)} - \underline{U}_i^{(\nu)}\right) = \underline{U}_i^{(\nu)} + \alpha \times \Delta\underline{U}_i^{(\nu+1)} \tag{11.55}$$

The value of the acceleration factor may have a substantial impact on the convergence of the iterative procedure. The optimal value of this factor, corresponding to a minimal number of iterations, depends on the configuration of the considered power system, on the parameters associated to the different components of the power system as well as on the level of burden of these components. Thus, the optimal value of the acceleration factor cannot be determined in advance. However, several suggestions in terms of guidance can be derived on the basis of experience. In that sense, the value of the acceleration factor $\alpha \in R$ should be between 1.1 and 1.8. It is worth to mention that the iterative procedure converges faster for the values of α larger than its optimal value than for those values that are smaller than its optimal one. In the cases where the optimal value of α is not known, it is preferable to use lower values from the specified interval. Inadequately chosen value for α can be a reason for the iterative process to converge slower or even diverge.

11.4.1 General Characteristics of the Gauss–Seidel Method

The Gauss–Seidel method is relatively simpler than the Newton–Raphson method and requires less computer memory. On the other hand, the number of iterations required for achieving convergence with predefined accuracy is relatively high and depends on the topology of the power system as well as on the parameters associated to the different components of the power system and the values of the injected power at different buses. It is not uncommon for this method to diverge. Besides that, the Gauss–Seidel method often diverges if there are components with negative reactance comprised in the power system.

From numerical algebra, it is known that the Gauss–Seidel method is characterized with worse convergence if the diagonal elements of the coefficient matrix of the system, which is obtained out of the system admittance matrix $[\underline{Y}]$ when omitting the row and the column corresponding to the slack bus, are not dominant. This is characteristic for those admittance matrices in which there is relatively high number of nodes that have only one connection with the rest of the system.

Additionally, if the number of loops in the graph of the power system is quite less than the number of nodes, that kind of power system is referred to as weakly connected. The specified disadvantages of the Gauss–Seidel method are especially emphasized in the case of large weakly connected power systems.

11.5 Direct Current Power Flow Method

The approximate direct current power flow model is obtained from the alternating current power system model, approximating that voltages in all buses are equal to nominal, considering the differences of voltage angles are very small and neglecting the losses in power system. The direct current power flow model gives a linear relationship between the power flows through the lines and the power injection at the nodes.

The direct current power flow model greatly simplifies the calculations by making a number of approximations including:

- Exclusion of the reactive power balance equations from analysis.
- Assumption that all substations voltages are equal to one per unit.
- The losses of the interconnections are neglected.

The named approximations presented as equations will be:

$$|U_i| \approx 1pu; \quad i = 1, 2, \ldots, N \tag{11.56}$$

where $|U_i|$ is the magnitude of the voltage in the bus i.

Approximation given by Eq. 11.56 states that the magnitude voltage in all buses in the analyzed system equals to nominal voltage and corresponds well to normal regime balanced networks.

$$\sin(\theta_i - \theta_k) \approx \theta_i - \theta_k; \quad i = 1, 2, \ldots, N; \quad k = 1, 2, \ldots, N \tag{11.57}$$

$$\cos(\theta_i - \theta_k) \approx 1; \quad i = 1, 2, \ldots, N; \quad k = 1, 2, \ldots, N \tag{11.58}$$

where θ_i is the phase angle of the voltage in the bus i.

The last two equations state that difference in voltage angles in high-voltage networks is very small and this approximation corresponds to the state in the real systems [1, 6].

11.6 Fast Decoupled Power Flow Method

Fast decoupled power flow method is one of the most extreme versions of approximated Newton–Raphson method. The method exploits the loose physical interaction between active (MW) and reactive (MVAR) flows in a power system by mathematically decoupling the MW-Θ and MVAR-V calculations [19–25].

Let us denote the complex power mismatch at bus k with $\Delta P_k + j\Delta Q_k$.

$$\Delta P_k = P_{k,\mathrm{sch}} - U_k \times \sum_{m\in k} U_m (G_{km} \times \cos\theta_{km} + B_{km} \times \sin\theta_{km}) \quad (11.59)$$

$$\Delta Q_k = Q_{k,\mathrm{sch}} - U_k \times \sum_{m\in k} U_m (G_{km} \times \sin\theta_{km} - B_{km} \times \cos\theta_{km}) \quad (11.60)$$

where $P_{k,\mathrm{sch}} + jQ_{k,\mathrm{sch}}$ is the scheduled complex power at bus k; θ_k and U_k are the voltage angle and magnitude at bus k, respectively; $\theta_{km} = \theta_k - \theta_m$ is the voltage angle; $G_{km} + jB_{km}$ is the (k, m)th element of bus admittance matrix $[Y]$; $\Delta\theta$ and ΔU are the voltage angle and magnitude corrections; $m \in k$ signifies that bus m is connected to bus k, including the case $m = k$; max $|\Delta P|$ and max $|\Delta Q|$ are the largest absolute element of the correction matrices $[\Delta P]$ and $[\Delta Q]$; and $\max|e_U|$ and $\max|e_S|$ are the largest absolute bus voltage magnitude error and branch MVA-flow error.

The polar form of the Newton–Raphson method is taken as a convenient starting point for the derivation. The Newton–Raphson method is the formal application of a general algorithm for solving nonlinear equations, and constitutes successive solutions of the sparse real Jacobian-matrix (see Eq. 11.32).

$$\begin{bmatrix} \Delta P \\ \Delta Q \end{bmatrix} = \begin{bmatrix} H & N \\ M & L \end{bmatrix} \times \begin{bmatrix} \Delta\theta \\ \Delta U/U \end{bmatrix} \quad (11.61)$$

The first step in applying the MW-Θ/MVAR-V decoupling principle is to neglect the coupling submatrices $[N]$ and $[M]$ in Eq. 11.61, giving two separated equations:

$$[\Delta P] = [H] \times [\Delta\theta] \quad (11.62)$$

$$[\Delta Q] = [L] \times \left[\frac{\Delta U}{U}\right] \quad (11.63)$$

where

$$H_{km} = L_{km} = U_k U_m (G_{km} \times \sin\theta_{km} - B_{km} \times \cos\theta_{km}) \quad \text{for } m \neq k \quad (11.64)$$

$$H_{kk} = -B_{kk} \times U_k^2 - Q_k \quad \text{and} \quad L_{kk} = -B_{kk} \times U_k^2 + Q_k \quad (11.65)$$

Equations 11.62 and 11.63 may be solved alternately as a decoupled Newton–Raphson method, re-evaluating and re-triangulating $[H]$ and $[L]$ at each iteration, but further physically justifiable simplifications may be made. In practical power systems, the following assumptions are almost always valid:

$$\cos\theta_{km} \approx 1; \quad G_{km} \times \sin\theta_{km} \ll B_{km}; \quad Q_k \ll B_{kk} \times U_k^2 \quad (11.66)$$

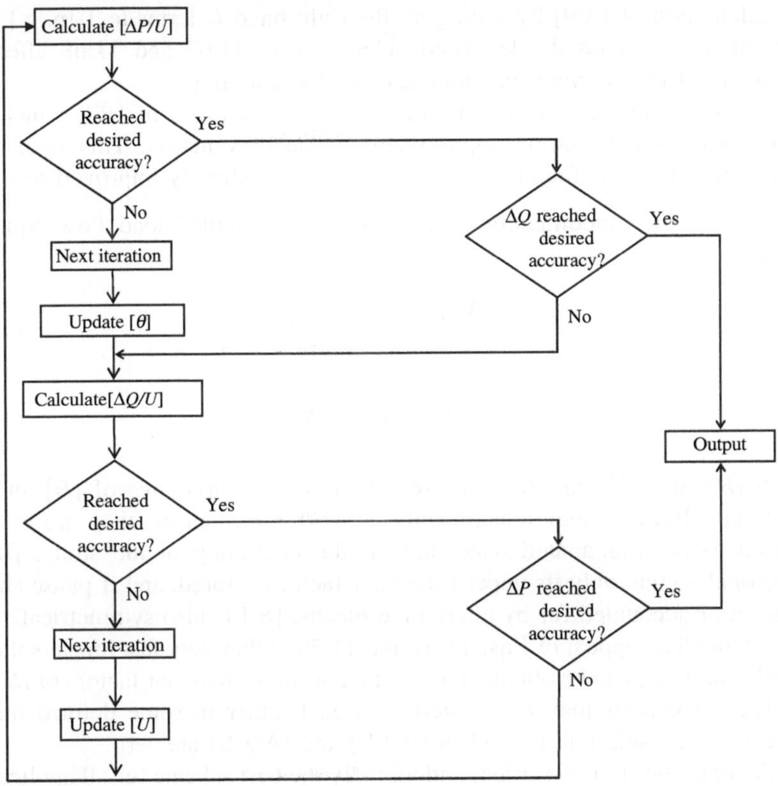

Fig. 11.6 Fast decoupled load flow method

so that good approximations to Eqs. 11.62 and 11.63 are:

$$[\Delta P] = \left[U \times B' \times U^T\right] \times [\Delta\theta] \tag{11.67}$$

$$[\Delta Q] = \left[U \times B'' \times U^T\right] \times \left[\frac{\Delta U}{U}\right] \tag{11.68}$$

At this stage of the derivation, the elements of the matrices $[B']$ and $[B'']$ are strictly elements of $[-B]$. The decoupling process and the final algorithmic forms are now completed by:

- Omitting from $[B']$ the representation of those network elements that predominantly affect MVAR flows, i.e., shunt reactances and off-nominal in-phase transformer taps.
- Omitting from $[B'']$ the angle-shifting effects of phase shifters.
- Taking the left-hand U terms in Eqs. 11.67 and 11.68 on to the left-hand sides of the equations, and then in Eq. 11.67 removing the influence of MVAR flows on

the calculation of $[\Delta\Theta]$ by setting all the right-hand U terms to 1 [p. u.]. Note that the U terms on the left-hand sides in Eqs. 11.67 and 11.68 affect the behaviors of the defining functions and not the coupling.

- Neglecting series resistances in calculating the elements of $[B']$, which then becomes the direct current approximation load flow matrix. This is of minor importance, but it is found experimentally to give slightly improved results.

With the above modifications, the final fast decoupled load flow equations become

$$\left[\frac{\Delta P}{U}\right] = [B'] \times [\Delta\theta] \tag{11.69}$$

$$\left[\frac{\Delta Q}{U}\right] = [B''] \times [\Delta U] \tag{11.70}$$

Both $[B']$ and $[B'']$ are real, sparse, and have the structures of $[H]$ and $[L]$, respectively. Because they contain only network admittances, they are constant and need to be triangulated once only at the beginning of the study. $[B'']$ is symmetrical so that only its upper triangular factor is stored, and if phase shifters are absent or accounted for by alternative means, $[B']$ is also symmetrical.

The immediate appeal of Eqs. 11.69 and 11.70 is that very fast repeat solutions for $[\Delta\Theta]$ and $[\Delta U]$ can be obtained using the constant triangular factors of $[B']$ and $[B'']$. These solutions may be iterated with each other in some defined manner toward the exact solution, i.e., when $[\Delta P/U]$ and $[\Delta Q/U]$ are zero.

Of the iteration strategies tried, undoubtedly the best scheme for all applications is to solve Eqs. 11.69 and 11.70 alternately, always using the most recent voltage values. Each iteration cycle comprises one solution for $[\Delta\Theta]$ to update $[\Theta]$ and then one solution for $[\Delta U]$ to update $[U]$, termed here the $(1\theta, 1U)$ scheme. Separate convergence tests are used for Eqs. 11.69 and 11.70 with the criteria:

$$\max|\Delta P| \le c_p, \quad \max|\Delta Q| \le c_q \tag{11.71}$$

where c_p and c_q are the convergence tolerances. Figure 11.6 shows the flow diagram of the process.

11.6.1 A General-Purpose Version of the Fast Decoupled Load Flow Method

This section briefly addresses one of the variations of the fast decoupled load flow method [18]. The key difference lies in the different way in which the resistances are ignored and in a different iteration scheme.

The standard fast decoupled load flow method is an alternative to the Newton–Raphson method, provided that three conditions are met: First, the voltages are around their nominal values; second, the angle differences across the lines are small; and third, the R/X ratios are small for all branches.

The first and the second condition are a serious problem only in a very small number of cases. But the third condition is important [22]. It prevents the use of the fast decoupled load flow method in systems where a small number of branches have relatively high resistances or where the overall R/X ratio is not small, which may be the case for low-voltage systems.

The intention to extend the use of the fast decoupled load flow method to systems with high R/X ratios can be realized by series or parallel compensation. Problematic branches are replaced by more branches each with a sufficient low R/X ratio. The structure is changed and the network size is expanded [23].

The method given in [24] presents an adaptation of the $[B']$ matrix, which is found solely experimentally. Cases with compensation can be solved. Cases without compensation can be solved with a slightly higher number of iterations.

A general-purpose version of the fast decoupled load flow method differs from the standard fast decoupled load flow method in two points:

- The handling of the resistances when building the $[B']$ and $[B'']$ matrices
- The iteration scheme that is used

The derivation procedure starts with decoupling the linearized load flow equations. The needed assumptions are the following:

- The resistances of the branches are small regarding their respective reactances.
- The angle differences are small.
- Several voltage magnitudes are set to 1 [p. u.], others are taken to the right-hand sides.
- The influence of the phase shifting of the phase shifters is ignored for building the $[B'']$ matrix.
- The influence of the off-nominal tap ratios of the transformers is ignored while forming the $[B']$ matrix.

The Eqs. 11.69 and 11.70 represent the system of interest to be solved. Both load flow matrices, $[B']$ and $[B'']$, are derived from the Jacobian matrix. They are formed out of the negation of the imaginary part of the admittance matrix, where:

- Shunts are omitted while forming $[B']$ matrix and are doubled while forming $[B'']$ matrix.
- The influence of the phase shifters is ignored while forming the $[B'']$ matrix.
- The influence of the off-nominal tap ratios is ignored while forming the $[B']$ matrix.

Both load flow matrices are built from network elements. In standard fast decoupled load flow method, which has excellent convergence properties, the resistances are ignored in the $[B']$ matrix only so this matrix is made of the branch reactances (variant XB).

In the general-purpose version of the fast decoupled load flow method, the resistances are ignored in the $[B'']$ matrix (variant BX). The number of iterations will be like that of the XB variant, but for systems with a few or with general high R/X ratios, the number of iterations needed to solve the load flow is considerably smaller than the number of the variant XB.

The possible other variants are variant BB and variant XX.

In variant BB, the resistances are not ignored at all and the branch susceptances are used for both $[B']$ and $[B'']$ matrices. This form of the fast decoupled load flow method usually suffers from a bad convergence.

In variant XX, the resistances are ignored in both $[B']$ and $[B'']$ matrices.

11.6.2 General Characteristics of the Fast Decoupled Load Flow Method

The fast decoupled load flow method combines many of the advantages of the existing methods. The algorithm is simpler, faster, and more reliable than Newton–Raphson method, and has lower storage requirements. The method is equally suitable for routine accurate load flows as for outage-contingency evaluation studies performed on or off-line.

11.7 Probabilistic Load Flow Method

The probabilistic load flow was first proposed in 1974 and has been further developed and applied into power system normal operation, short-term/long-term planning as well as other areas [26–32]. The probabilistic load flow requires inputs with probability density function to obtain system states and power flows in terms of probability density function, so that the system uncertainties can be included and reflected in the outcome. The probabilistic load flow can be solved numerically, i.e., using a Monte Carlo method, or analytically, e.g., using a convolution method, or a combination of them [30–32].

11.7.1 Numerical Probabilistic Load Flow

The probabilistic load flow is in principle doing deterministic load flow for a large number of times with inputs of different combinations of nodal power values. The exact nonlinear form of the load flow equations, Eqs. 11.13, 11.14, 11.17, and 11.18, can be used in the probabilistic load flow analysis. The random number generation and random sampling are used for determining the nodal power values.

The higher is the number of the combinations of the nodal values, the more accurate the load flow calculation is.

11.7.2 Analytical Probabilistic Load Flow

This section briefly addresses the general probabilistic load flow analytical method [26]. The analytical probabilistic load flow is an arithmetic method using convolution techniques with probability density functions of stochastic variables of power inputs, so the probability density functions of stochastic variables of system states and line flows can be obtained. The objective is to find a set of corresponding values of branch flows for a power system with constant configuration, line parameters, and a given set of probable values of node loads. The uncertainty of load data can depend on measurement error or forecast inaccuracy, on the assumption of load within certain limits, and on unscheduled outage. The load is not known accurately, but a range of values is given together with their frequency of occurrence instead.

The proposed method assumes that the loads are static by considering the condition over a small time interval and therefore the loads are random variables [26]. The branch flows in the network are a function of loads. The loads and the branch flows are random variables.

The probability model for load flow has the following advantages:

- All power inputs or outputs can be given as a set of values.
- It does not exclude the conventional load flow calculations.
- Degrees of importance or frequency of occurrence of a given load data can be respected by associated probability.
- The synthesis of all possible branch flows can be obtained in the form of distribution functions of branch flows.

The probability model for load flow has the following disadvantages:

- The nonlinear relation between the node loads and branch flows.
- Because the generation has to meet the demand plus losses, the mathematical model of the probabilistic load flow must take into account the control of the balance of power.
- The mathematical control of the balance is nonlinear complicated function of the power inputs and outputs in particular nodes.
- The number of data processed is much greater than in the conventional load flow; so suitable numerical method is needed.

It is therefore desirable to make the simplified assumptions. The more precise formulation of the method includes:

- The graph of the *N*-nodes, *B*-branches, network, and the parameters of the branches; the probability of this graph is equal to one.

- R distribution functions of the real power inputs and outputs; ($R \times N$).
- The rule or procedure of balancing the power in the case of surplus or deficiency; the rule can be given analytically or in the form of an algorithm.

The following assumptions are made to solve the distribution functions of branch flows:

- Branch flows are linearly related to net nodal loads.
- Active and reactive power flows are independent of each other.
- The balancing of power is a function of the sum of power inputs and outputs only, and it is not dependent on the power inputs and outputs in particular nodes.

The detailed mathematical model is presented in the reference [26].

11.7.3 General Characteristics of the Probabilistic Load Flow Method

The probabilistic treatment of loads in a load flow has many applications, particularly when implemented in an alternating current load flow.

The numerical approach, e.g., a Monte Carlo method, substitutes a chosen number of values for the stochastic variables and parameters of the system model and performs a deterministic analysis for each value so that the same number of values are obtained in the results; whereas the analytical approach analyzes a system and its inputs using mathematical expressions, e.g., probability density functions, and obtains results also in terms of mathematical expressions. The main concern about the Monte Carlo method is the need of large number of simulations, which is very time-consuming, whereas the main concerns about the analytical approach are the complicated mathematical computation and the accuracy because of different approximations.

The general analytical approach [26] for solving a probabilistic load flow was considered. The presented method enables the evaluation of the expected values, standard deviations, and distribution functions of branch flows when the configuration and parameters of the network are constant and the power inputs and outputs are random variables. The given data may come from statistical records or can be a set of arbitrary values. In the first case, results of calculations will give the probability of occurrence and the second case will give the synthesis of all possible branch flows corresponding to the given data.

The described method of calculation can be used for planning and operational purposes. The results of calculations provide much more information about the load condition of a network than the information obtained from conventional load flow method. The density function, apart from the expected value and standard deviation, gives the answer to some questions, which are important from the practical viewpoint, such as:

- What is the probability that the branch flow will exceed the capacity limit or will be greater or less than a certain value?
- What percentage of all possible values of branch load belongs to the economically desirable range of branch load values?
- What is the practically possible range of branch load values?
- What is the most probable load value?

Generally, the method can be helpful in all problems in which the load conditions of a network should be analyzed by uncertainty or variety of data available. In particular, it can be applied to the solution of the following problems:

- The proper choice of the number, capacity, and the configuration of the branch in a network.
- The evaluation of operational cost and economical effectiveness of a network; the systematic error in calculation of network losses can be eliminated using probabilistic load flow.
- The forecasting of load inputs and outputs in the network planning and operation taking into account forecasting error.
- The assessment of reliability of power supply in the particular network substations.

References

1. Grainger JJ, Stevenson WD (1994) Power system analysis. McGraw-Hill, New York
2. Rajičić D, Taleski R (1996) Methods for analysis of power systems (in Macedonian). Faculty of Electrical Engineering, Skopje
3. Tinney WF, Hart CE (1967) Power flow solution by Newton's method. IEEE Trans Power App Syst 86:1449–1460
4. Stott B (1971) Effective starting process for Newton–Raphson load flow. Proc Inst Electr Eng 118:983–987
5. Treece JA (1969) Bootstrap Gauss–Seidel load flow. Proc Inst Electr Eng 116(5):866–870
6. Volkanovski A (2008) Impact of offsite power system reliability on nuclear power plant safety. PhD thesis, University of Ljubljana
7. Deckmann S, Pizzolante AC, Monticelli AJ et al (1999) Numerical testing of power system load flow equivalents. IEEE Trans Power App Syst 6:2292–2300
8. Deckmann S, Pizzolante AC, Monticelli AJ et al (1999) Studies on power system load flow equivalents. IEEE Trans Power App Syst 6:2301–2310
9. Stott B (1974) Review of Load-Flow Calculation Methods, Proc Inst Electr Eng 62(7)
10. Stott B (1971) Effective starting process for Newton-Raphson load flows. Proc Inst Electr Eng 118(8):983–987
11. Mori H, Tanaka H, Kanno J (1996) A preconditioned fast decoupled power flow method for contingency screening. IEEE Trans Power Syst 11(1):357–363
12. Alves AB, Asada EN, Monticelli A (1999) Critical evaluation of direct and iterative methods for solving ax=b systems in power flow calculations and contingency analysis. IEEE Trans Power Syst 12(4):702–708
13. Wood AJ, Wollenberg BF (1996) Power Generation, Operation and Control. Wiley,
14. Powell L (2004) Power System Load Flow Analysis, McGraw-Hill Professional Series

15. Glimn AF, Stagg GW (1957) Automatic Calculation of Load Flows, AIEE Summer General Meeting, pp 24–28
16. Stott B, Jardim J, Alsaç O (2009) DC Power Flow Revisited. IEEE Trans Power Syst 24(3)
17. Srinivas MS (2000) Distribution load flows: a brief review, Power Engineering Society Winter Meeting, IEEE, pp 942–945
18. van Amerongen RAM (1989) A General-Purpose Version of the Fast Decoupled Load Flow. IEEE Trans Power Syst 4(2):760–770
19. Stott B, Alsac O (1974) Fast Decoupled Load Flow. IEEE Trans Power App Syst 93(3): 859–869
20. Peterson NM, Tinney WF, Bree DW (1972) Iterative Linear AC Power Flow Solution for Fast Approximate Outage Studies. IEEE Trans Power App Syst 91:2048–2053
21. Tinney WF, Peterson NM (1971) Steady State Security Monitoring, Proc Symposium on Real Time Control of Electric Power Systems, Brown, Boveri & Comp Ltd, Baden
22. Wu FF (1977) Theoretical Study of the Convergence of the Fast Decoupled Loadflow. IEEE Trans Power App Syst 96:268–275
23. Haley PH, Ayres M (1985) Super Decoupled Loadflow with Distributed Slack Bus. IEEE Trans Power App Syst 104:104–113
24. Rajičić D, Bose A (1987) A Modification to the Fast Decoupled Power Flow for Networks with high R/X ratios, Proceedings of PICA Conference, pp 360–363
25. Rajičić D, Bose A (1988) A modification to the Fast Decoupled Power Flow for Networks with high R/X ratios. IEEE Trans Power Syst 3(2):743–746
26. Borkowska B (1974) Probabilistic load flow. IEEE Trans Power App Syst 93(3):752–755
27. Allan RN, Borkowska B, Grigg CH (1974) Probabilistic Analysis of Power Flows. Proc Inst Electr Eng 121(12):1551–1556
28. Allan RN, Leite da Silva AM, Burchett RC (1981) Evaluation methods and accuracy in probabilistic load flow solutions. IEEE Trans Power App Syst 100(5):2539–2546
29. Leite da Silva AM, Ribeiro SMP, Arienti VL et al (1990) Probabilistic load flow techniques applied to power system expansion planning. IEEE Trans Power Syst 5(4):1047–1053
30. Jorgensen P, Christensen JS, Tande JO (1998) Probabilistic load flow calculation using Monte Carlo techniques for distribution network with wind turbines. In: Proceedings of the 8th international conference on harmonics and quality of power 2, pp 1146–1151
31. Allan RN, Grigg CH, Al-Shakarchi MRG (1976) Numerical techniques in probabilistic load flow problems. Int J Num Methods Eng 10:853–860
32. Leite da Silva AM, Arienti VL (1990) Probabilistic load flow by a multilinear simulation algorithm. Proc Inst Electr Eng Part C 137(4):276–282
33. Su CL (2005) Probabilistic load-flow computation using point estimate method. IEEE Trans Power Syst 20(4):1843–1851
34. Ačkovski R (1989) Contribution on methods for planning and development of power systems using Monte Carlo simulation. PhD thesis, Faculty of Electrical Engineering, Skopje
35. Todorovski M (1995) Approximate calculation of power flows thought high voltage network. Thesis, Faculty of Electrical Engineering, Skopje
36. Zhu J (2009) Optimization of power system operation. Wiley, Piscataway, NJ

Part IV
Reliability of Power Systems

Chapter 12
Generating Capacity Methods

Power is not revealed by striking hard or often, but by striking true

Honoré de Balzac

12.1 Introduction

The primary function of a power system is to provide electrical energy to its customers as economically as possible with an acceptable degree of quality [1–5]. Reliability of power supply is one of the features of power quality [6–8]. The two constraints of economics and reliability are competitive because increased reliability of supply generally requires increased capital investment. These two constraints are balanced in many different ways in different countries and by different utilities, although generally they are all based on various sets of criteria.

A wide range of related measures or indicators can be determined using probability theory. A single all-purpose formula or technique does not exist. The approaches and their respective mathematical expressions depend on the defined problem and determined assumptions. Several assumptions must be made in practical applications of probability and statistical theory. The validity of the analysis is directly related to the validity of the model used to represent the system. Actual failure distributions rarely completely fit the analytical descriptions used in the analysis, and care must be taken to ensure that significant errors are not introduced through oversimplification of a problem.

The most important aspect of good modeling and analysis is to have a complete understanding of the engineering implications of the system. No amount of probability theory can circumvent this important engineering aspect.

There are two main categories of evaluation techniques: (i) analytical and (ii) simulation. Analytical techniques represent the system by a mathematical model and evaluate the measures or indicators from this model using mathematical solutions. Simulation techniques estimate the measures or indicators by simulating the actual process and random behavior of the system. Probabilistic simulation techniques are a subset of simulation techniques that treat the problem as a series of real experiments. The input parameters of each simulated experiments are obtained using Monte Carlo selection of their values. Both categories of methods have advantages and disadvantages [9–19]. The Monte Carlo simulation requires a large amount of

M. Čepin, *Assessment of Power System Reliability*,
DOI: 10.1007/978-0-85729-688-7_12, © Springer-Verlag London Limited 2011

computing time and is not used extensively if alternative analytical methods are available. On the other side, if the analytical methods are too complex, the probabilistic simulations can give a good approximation of results. The resulted measures or indicators in both categories of evaluation techniques are only as good as the model derived for the system, the appropriateness of the evaluation technique, and the quality of the assumptions and input data used in the models.

12.2 Review of Indicators Considering Loss of Power

12.2.1 Generation Reserve Margin

The installed capacity in power system must be higher than expected consumption. A reserve power needs to be provided for frequency regulation and for case of major aggregate loss of capacity. Generation reserve margin is a measure that shows how the capacity of power system exceeds the peak consumption. Generation reserve margin is defined as [6]:

$$\text{Generation reserve margin (\%)} = \frac{\text{Capacity in service} - \text{Peak load}}{\text{Capacity in service}} \times 100 \tag{12.1}$$

Cost of an interruption is significant. The interruption duration cost is not necessarily a linear function of duration. Reported cost of an interruption of providing energy to the consumer is approximately 100 times higher than the average price of electric energy [6].

By power system planning, loss of generating capacities must be considered with a fact that consumption may not be covered the whole operating time. Lack of production is shown with a risk degree.

12.2.2 Percent Reserve Evaluation

The earliest method and most easily computed criterion for evaluation of generation system adequacy is the percent generation reserve margin approach. This method is sensitive to only two factors at one point in time.

Percent reserve evaluation computes the generation capacity exceeding annual peak load. It is calculated by comparing the total installed generating capacity at peak with the peak load. The criterion is based on past experience requiring reserve margins in the range of 15–25% to meet demand. Satisfactorily meeting load demand meant that the frequency and magnitude of emergency power purchases from neighboring power systems were reasonable and/or the number of curtailments was small.

There are, however, disadvantages of the percent reserves approach. It is insensitive to forced outage rates and unit size considerations, power transfer capacity, and failures in transmission network as well as to differing load characteristics of power systems. Although this approach is a useful step in the analysis of generation reserve problems, it does not provide a complete answer to how much generation capacity is required to adequately serve load demands.

12.2.3 Loss of the Largest Generating Unit Method

Loss of the largest generating unit method provides a degree of sophistication over the percent reserve margin method by reflecting the effect of unit size on reserve requirements. With the loss of the largest unit method, required reserve margin is calculated by adding the size of the largest unit divided by the peak load plus a constant reserve value.

For example, if reserve requirements are 15% plus the largest unit, and the largest unit is 500 MW in a power system with a 5,000-MW peak load, then the reserve requirement is 15% + 500/5,000×(100%), or 25%. This approach begins to explicitly recognize the impact of a single outage, that is, loss of the largest generating unit. Probabilistic measures are necessary to extend this method to include multiple simultaneous outages.

Loss of the largest unit method, although simple, has a distinct advantage over the generation reserve margin method. As larger units are added to a system, the percent reserves for a system are implicitly increased by this method as needed. But similarly as percent reserve evaluation method, it is insensitive to forced outage rates of the units and power transfer capacity and failures in the transmission network.

12.2.4 Static Analysis of Loss of Capacity

The basic probability principles and combining the different generating units are used for calculating the probability of aggregate loss of capacity. The data and the results are usually represented in capacity outage probability tables. The success probability or availability and its complement, i.e., failure probability or unavailability of each generating unit are the input data. All combinations of available and unavailable generating units are presented in tabular form together with the calculated system availability.

Example of a power system includes three generating units. First two generating units (A1 and A2) have an output of 30 MW each and the third one (A3) has the output of 70 MW. Unavailability of each generating unit is 0.02. Availability of each generating unit is calculated from unavailability and it is 0.98. Table 12.1 shows all possible combinations of available or unavailable generating units.

Table 12.1 All possible combinations of generated power

Power loss (MW)	A1 A2 A3	Cumulative success probability
0	A × A × A	0.941192
30	U × A × A	0.019208
30	A × U × A	0.019208
60	U × U × A	0.000392
70	A × A × U	0.019208
100	U × A × U	0.000392
100	A × U × U	0.000392
130	U × U × U	0.000008

12.3 Loss of Load Probability

A loss of load probability (LOLP) is a probabilistic approach for determination of required reserves, which was developed in the year 1947 [1]. This approach examines the probabilities of simultaneous outages of generating units that, together with a model of daily peak-hour loads, determine the number of days per year of expected capacity shortages. Today, LOLP is the most widely accepted approach in the utility industry for evaluating generation capacity requirements [6].

Loss of load occurs whenever the system load exceeds the available generating capacity. The LOLP is defined as the probability of the system load exceeding available generating capacity under the assumption that the peak load is considered as constant through the day.

The loss of load probability does not really stand for a probability. It expresses statistically calculated value representing the percentage of hours or days in a certain time frame, when energy consumption cannot be covered considering the probability of losses of generating units. This time frame is usually 1 year, which can be represented as 100% of time frame. In other words, the LOLP stands for an expected percentage of hours or days per year of capacity shortage. The LOLP actually does not stand for a loss of load but rather for a deficiency of installed available capacity.

The term LOLP is closely related to the term loss of load expectation (LOLE), which is presented in next section. If the time interval used for the LOLP is expressed in the time units instead in percentage values, the LOLE is obtained instead of the LOLP.

The generation system planners can evaluate generation system reliability and determine how much capacity is required to obtain a specified level of LOLP. As demand grows over time, additional generating units are included in a way that the LOLP does not exceed the required criterion. LOLP usually varies exponentially with load changes.

While the effect of random outages is evaluated probabilistically, scheduled outages are evaluated deterministically. Deterministic risk criteria such as percentage reserve and loss of target unit do not define consistently the true risk in the system.

12.3.1 Loss of Load Probability Definition

Loss of one generating unit causes the expected risk of loss of power supply $E(t)$, which is also known as mathematical expectation and is defined as:

$$E_i(t) = p_i t_i \tag{12.2}$$

where p_i is the probability of loss of capacity, and t_i is the duration of loss of capacity in percent.

Loss of load probability for the whole system is defined as a sum of all mathematical expectations for all units:

$$\mathrm{LOLP} = \sum_{i=1}^{n} p_i t_i \tag{12.3}$$

12.3.2 Loss of Load Probability During Scheduled Outages

The planner of power generation must schedule planned outages during the year, because the generating units must be regularly maintained and inspected.

Short-term maintenance process is continually updated. If a generating unit experiences a long-forced outage, the annual maintenance schedule for the power system can be reshuffled to further improve system reliability and to decrease the power system production costs.

Planned outage requirements of power plants usually have a cyclical pattern. The maintenance procedure schedules the maintenance of generating units, so that available generation capacity reserve is the same for all weeks. This kind of procedure has the lowest LOLP.

The most widely used algorithm for scheduling maintenance consists of four steps [6]:

- Arrange generating units by size with the largest unit first and the smallest unit last.
- Schedule the largest generating unit for maintenance during periods of the lowest load.
- Adjust weekly peak load by the generating unit capacity on maintenance.
- Repeat the second and the third step until all generating units are scheduled for maintenance.

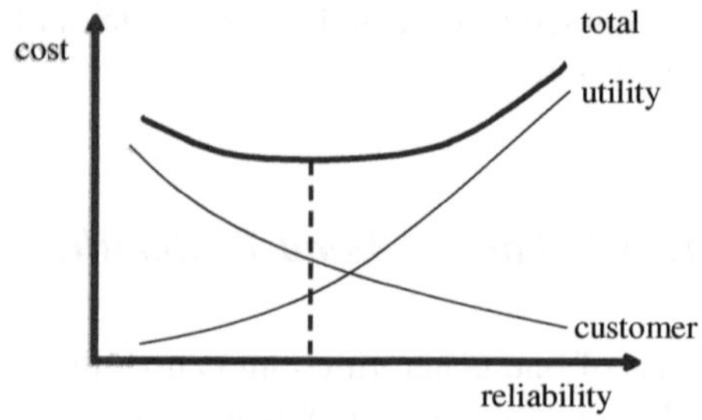

Fig. 12.1 Costs of power system and its reliability

12.3.3 Loss of Load Probability Annual Calculations

The annual calculations of the loss of load probability are performed in four steps [6]:

- Computing the annual maintenance scheduled of generation units.
- Building the capacity outage table using only the capacity available for service during particular week.
- Computing of daily outage probabilities and accumulating in a weekly index.
- The process can be repeated for each week in the year.

12.3.4 Loss of Load Probability Optimum Reliability Level

One approach to calculate LOLP optimum reliability level is to base the design target on a historical review.

Another approach is an analytical one. The total costs of electricity for several different levels of the LOLP index are calculated. The LOLP level that provides the lowest total costs is selected. The procedure consists of four steps [6]:

- Determine utility cost to improve reliability.
- Determine cost saving to electrical energy customers for improved reliability.
- Compute total cost as the sum of steps 1 and 2.
- Find the minimum cost by repeating steps 1, 2, and 3 alternate reliability levels.

The total costs include the costs of utility and costs to the consumer. Figure 12.1 shows the costs of power system.

12.3.5 Loss of Load Probability Calculation

The easiest approach for calculation of LOLP is to represent the generation system with a state enumeration table. State enumeration table is a table, where the combinations of available and unavailable generating units are ordered in rows and

Table 12.2 Generation system data

Generating unit	Capacity (MW)	Unit failure probability	Unit success probability
Unit A	50	0.05	0.95
Unit B	100	0.07	0.93
Unit C	200	0.10	0.90
System includes units A, B, and C	350		

the respective availabilities and unavailabilities of the generating units are considered for the system availability and unavailability calculation. The combinations of available and unavailable generating units are expressed in terms of probability, where the product of availability of working units and unavailability of units in outage gives the probability of respective combination. The state enumeration table can be arranged in monotonically increasing order of combinations of available and unavailable generating units in terms of increasing overall power of considered generating units.

The state enumeration table is further presented as a cumulative outage table for assessment of the probability of not being able to supply the necessary capacity. The LOLP is obtained through multiplications of the obtained probabilities with the time intervals for the corresponding states and through the sum of those products.

12.3.6 Loss of Load Probability Example

Table 12.2 shows the example generation system data. The generation system is composed of three generating units: A, B, and C, with their power capacity of 50, 100, and 200 MW, respectively. The failure probability of each unit is given and the success probability of each unit is given. The alternative for unit failure probability is the forced outage rate (FOR) or the unavailability of the unit. The alternative for unit success probability is the availability of the unit.

Eight combinations considering the generating units in outage or in service exist. Table 12.3 shows the outage state enumeration table, which enumerates all of these states and the probability of each. For example, the probability that no generating unit is in outage, i.e., all units A, B, and C are in service, is the product of the success probabilities of units A, B, and C. Similarly, another state may be that unit A is an outage and B and C are in service. The probability of this state is the product of the unit failure probability for unit A multiplied with unit success probabilities of units B and C.

Table 12.3 can be ordered in monotonically increasing order of overall capacity of units in outage as presented in Table 12.4.

Consider evaluating the probability of not being able to supply a 220 MW load demand. Because the capacity of the three unit system is 350 MW, the load could

Table 12.3 Outage state enumeration

Units in outage	Capacity in outage (MW)	Units in service	Probability
None	0	A, B, C	$0.95 \times 0.93 \times 0.90 = 0.79515$
A	50	B, C	$0.05 \times 0.93 \times 0.90 = 0.04185$
B	100	A, C	$0.95 \times 0.07 \times 0.90 = 0.05985$
C	200	A, B	$0.95 \times 0.93 \times 0.10 = 0.08835$
A, B	150	C	$0.05 \times 0.07 \times 0.90 = 0.00315$
A, C	250	B	$0.05 \times 0.93 \times 0.10 = 0.00465$
B, C	300	A	$0.95 \times 0.07 \times 0.10 = 0.00665$
A, B, C	350	None	$0.05 \times 0.07 \times 0.10 = 0.00035$
			Sum = 1

Table 12.4 Outage state enumeration based on monotonically increasing order

Capacity in outage (MW)	Capacity in service (MW)	Probability
0	350	0.79515
50	300	0.04185
100	250	0.05985
150	200	0.00315
200	150	0.08835
250	100	0.00465
300	50	0.00665
350	0	0.00035

not be supplied if capacity of 130 MW or more is on outage ($350 - 220 = 130$). The probability of 130 MW of more in outage is calculated as a cumulative probability according to data from Table 12.4:

$$0.00315 + 0.08835 + 0.00465 + 0.00665 + 0.00035 = 0.10315$$

Hence, the probability of not meeting load demand is 0.10315.

Because computation of the probability of not meeting the load demand requires an evaluation of the determined capacity in outage including larger values of capacity in outage, Table 12.4 can be written as a cumulative outage table as presented in Table 12.5. The first row of the table under the heading row with indicating capacity of 0 MW or more in outage is the sum of corresponding probabilities from Table 12.4, which represent capacities in outage from 0 to 350 MW. The second row with indicating capacity of 50 MW or more in outage represents the sum of corresponding probabilities from Table 12.4, which represent capacities in outage from 50 to 350 MW. The third row with indicating capacity of 100 MW or more in outage represents the sum of corresponding probabilities from Table 12.4, which represent capacities in outage from 100 to 350 MW. The other rows are obtained similarly.

Table 12.5 is expanded to Table 12.6 by adding the time intervals of certain capacities in outage and the contributions to LOLP. The LOLP is the sum of contributions from the right column of the Table 12.6. The LOLP for the example case is 0.19.

Table 12.5 Cumulative outage state enumeration

Capacity in outage (MW)	Probability of determined capacity or more in outage
0 or more	1
50 or more	0.20485
100 or more	0.16300
150 or more	0.10315
200 or more	0.10000
250 or more	0.01165
300 or more	0.00700
350 or more	0.00035

Table 12.6 Contributions to loss of load probability

Capacity in outage (MW)	Probability of determined capacity or more in outage	Time interval of capacity in outage (%)	Contributions to loss of load probability
0 or more	1	0	0
50 or more	0.20485	0	0
100 or more	0.16300	0	0
150 or more	0.10315	0	0
200 or more	0.10000	0	0
250 or more	0.01165	10	0.1165
300 or more	0.00700	10	0.0700
350 or more	0.00035	10	0.0035

12.4 Loss of Load Expectation

Loss of load expectation represents the probability that aggregates will not be able to cover the necessary power consumption. The term LOLE is closely related to the term LOLP. If the time interval used for the LOLP is expressed in the time units instead in percentage values, the LOLE is obtained instead of the LOLP. Figure 12.2 shows yearly load diagram [7].

The limit value of LOLE for a reliable supply is 10 h per year. In some European countries, the limit can also be settled between 4 and 8 h per year. Power system with a higher value of a LOLE has a lack of power charging or the existing units are badly disposable.

12.4.1 Loss of Load Expectation Definition

Loss of load expectation can be obtained using the daily peak load variation curve. A particular capacity outage contributes to the system by an amount equal to the product of the probability of existence of the particular outage and the number of time units. The period of study could be week, month or a year. The simplest

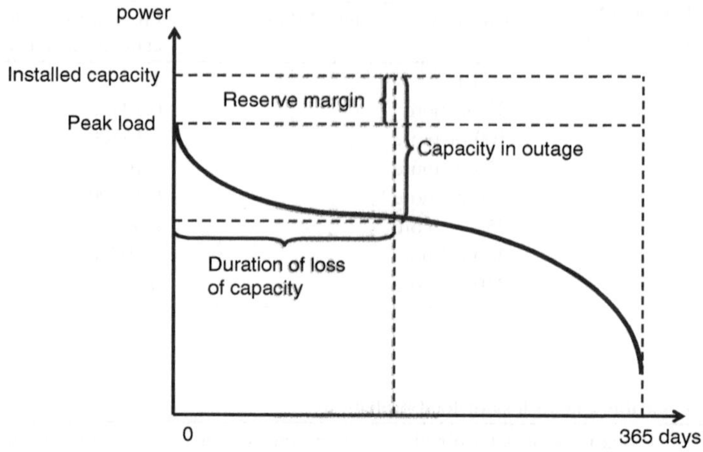

Fig. 12.2 Yearly load diagram

application is the use of the curve on yearly basis. When using a daily peak load variation curve on annual basic, the LOLE is in days per year.

$$\text{LOLE} = \sum_{i=1}^{n} p_i t_i \tag{12.4}$$

where p_i is the individual probability of capacity in outage and t_i is the duration of loss of power supply in days.

When the cumulative probability P_i is used, LOLE is defined as:

$$\text{LOLE} = \sum_{i=1}^{n} P_i (t_i - t_{i-1}) \tag{12.5}$$

LOLE is also defined with a probability that consumption L will not be covered during working power capacity C.

$$\text{LOLE} = \sum_{i=1}^{n} P_i (C_i - L_{i-1}) \tag{12.6}$$

12.4.2 Input Parameters

Input parameters for calculating the LOLE [1]:

- Consumed energy including losses
- Influence of hydrology on hydro power plant production
- Non-availability of coal power plants during scheduled outages

- Non-availability of coal power plants during random outages
- Import and export of electrical energy
- Limited load in power system

12.4.3 Evaluation Methods on Period Bases

The basic LOLE approach is very flexible. There are three ways in which the LOLE method can be used to determine an annual risk index:

- Monthly (or period) basis considering maintenance
- Annual basis neglecting maintenance
- Worst-period basis

12.4.3.1 Monthly Approach

The appropriate capacity outage probability table is combined with the corresponding load characteristic. If the capacity on maintenance is not constant during the month, the month can be divided into several intervals during which the capacity is constant. This method assumes that the monthly peak can occur on any day during the period. The total reliability measure is obtained by summing the interval values. The annual reliability measure is the sum of the 12 monthly reliability measures.

12.4.3.2 Annual Approach

The annual forecast peak and system load characteristic are combined with the system capacity outage probability table to give an annual reliability level. A constant capacity level must exist for the entire period. If the year can be divided into a peak-load season and a light-load season, the planned maintenance may be scheduled entirely in light-load season.

12.4.3.3 Worst-Period Approach

In some cases, the load level in a particular season or even in a month may be so high, that this value dominates the annual figure. A reliability criterion for such a system can be obtained using only the worst period value. If the December is the month with the highest monthly risk period, an annual risk figure can be obtained by multiplying the December value by 12.

Table 12.7 Generation units data

Generating unit	Capacity (MW)	Unit failure probability	Unit success probability
A	40	0.10	0.90
B	30	0.05	0.95
C	10	0.04	0.96

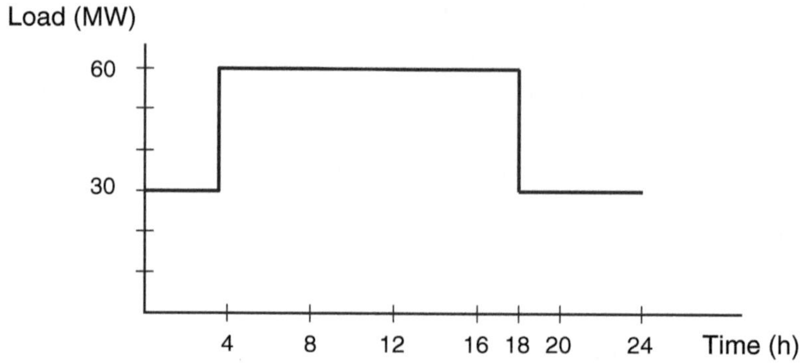

Fig. 12.3 Daily load diagram

12.4.4 Loss of Load Expectation Calculation

Also at LOLE calculation, the most convenient procedure is representing the power system in tables, including capacities, unavailabilities, and availabilities. Usually, a daily load diagram is given and it should be considered.

All unit outage combination probabilities are calculated and represented as cumulative probabilities. LOLE index is calculated as a product of expected load duration and cumulative outage probability for required state.

12.4.5 Loss of Load Expectation Example

The example generation system consists of three units. Table 12.7 shows the corresponding generation system data. Figure 12.3 shows a daily load diagram.

There are eight combinations of generating units in outage or in service. Mark 0 stands for outage of corresponding unit in the Table 12.8, mark 1 stands for operating unit in service.

The example on Fig. 12.3 with two load levels show 14 h of 60 MW between 4 and 18 h and 10 h of 30 MW between 0 and 4 h and between 18 and 24 h.

- *At 30 MW*: The applicable outage states for the LOLE are those outage states, where the capacity in service 30 MW is not reached, i.e., those outage states in the last two rows. The cumulative probability of those two states is 0.005. Therefore, the LOLE is calculated as $\text{LOLE}_1 = 0.005 \times 10\ \text{h} = 0.05\ \text{h}$.

Table 12.8 Outage states

Unit A	Unit B	Unit C	Capacity in outage (MW)	Capacity in service (MW)	Probability of each capacity	Cumulative Probability	
1	1	1	0	80	$0.9 \times 0.95 \times 0.96 = 0.8208$	1	
1	1	0	10	70	$0.90 \times 0.95 \times 0.04 = 0.0342$	0.1792	
1	0	1	30	50	$0.90 \times 0.05 \times 0.96 = 0.0432$	0.145	
0	1	1	40	40	$0.10 \times 0.95 \times 0.96 = 0.0912$	0.1018	0.1124
1	0	0	40	40	$0.90 \times 0.05 \times 0.04 = 0.0018$	0.0106	
0	1	0	50	30	$0.10 \times 0.95 \times 0.04 = 0.0038$	0.0088	
0	0	1	70	10	$0.10 \times 0.05 \times 0.96 = 0.0048$	0.005	
0	0	0	80	0	$0.01 \times 0.05 \times 0.04 = 0.0002$	0.0002	

- *At 60 MW*: The applicable outage states for the LOLE are those outage states, where the capacity in service 60 MW is not reached, i.e., those outage states in the last six rows. The cumulative probability of those two states is 0.145. Therefore, the LOLE is calculated as $\text{LOLE}_2 = 0.145 \times 14\ \text{h} = 2.03\ \text{h}$.
- LOLE for whole day is a sum of both: $\text{LOLE} = \text{LOLE}_1 + \text{LOLE}_2 = 0.05\ \text{h} + 2.03\ \text{h} = 2.08\ \text{h}$

To get the annual LOLE, the calculations have to be repeated for every day.

12.5 Review of Indicators Considering Loss of Energy

The area under the load duration curve represents the energy generated during the specific time interval and can be used to calculate an expected energy not supplied because of insufficient installed capacity (Fig. 12.4). The ratio between the energy curtailed because of reduced capacity because of specific capacity in outage and the total energy generated can be defined as energy index of unreliability. Its complement is energy index of reliability (EIR).

Any outage of generating capacity exceeding the reserve results in a curtailment of system load energy [7]. The probable energy curtailed is obtained as a product of a probability of magnitude of specific capacity in outage and the energy curtailed by a specific capacity in outage. The sum of those gives the loss of energy expectation (LOEE).

$$\text{LOEE} = \sum_{i=1}^{n} P_i E_i \tag{12.7}$$

where O_i is the magnitude of specific capacity in outage, P_i is the probability of magnitude of specific capacity in outage, E_i is the energy curtailed by a specific capacity in outage, and E is the total energy under the load duration curve.

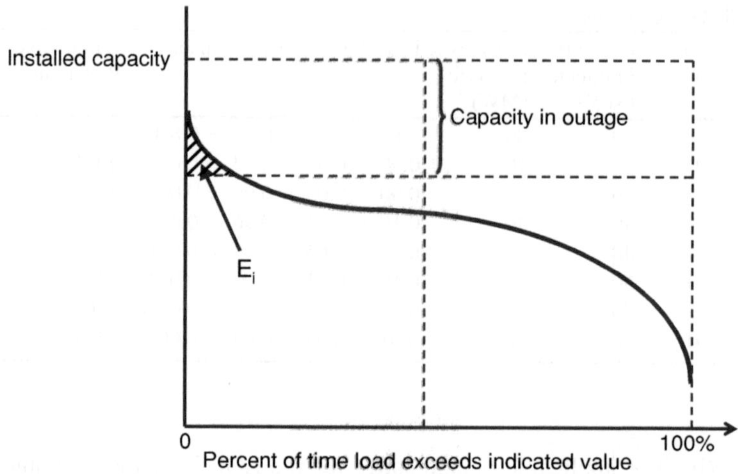

Fig. 12.4 Energy curtailment by a given capacity in outage or loss of energy expectation

Normalized loss of energy expectation ($LOEE_{\%}$ or $LOEE_{p.u.}$) is obtained by using the total energy under the load duration curve. Normalized loss of energy expectation equals to energy index of unreliability.

$$LOEE_{\%} = \sum_{i=1}^{n} \frac{P_i E_i}{E} \tag{12.8}$$

Energy index of reliability is obtained as its complement.

$$EIR = 1 - LOEE_{\%} \tag{12.9}$$

12.6 Frequency and Duration Method

Frequency and duration method is more complex than the LOLE, but it is also more complete method to evaluate the static capacity suitability for a given generation system [7–11]. The LOLE method gives neither any indication of the frequency of occurrence of an insufficient capacity condition nor the duration for which it is likely to exist and is not as sensitive to variations in the two individual elements, unit failure rate and unit repair rate, as the frequency and duration method. Frequency and duration are the most useful indices for customer or load point evaluation.

The frequency and duration method requires additional data to that in other methods. The LOLE method requires only steady state availability and unavailability parameters. The frequency and duration method uses the transition rate parameters μ and λ in addition to availability and unavailability. Parameter λ

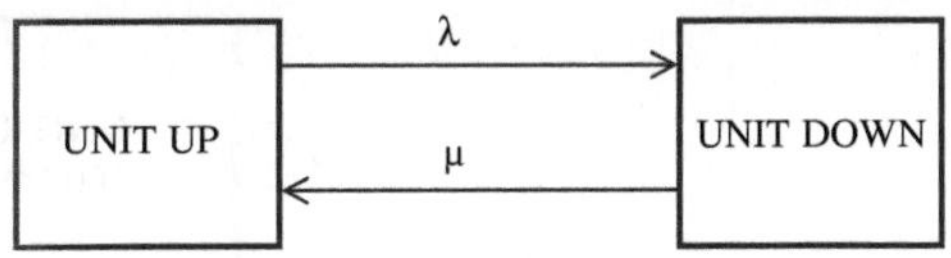

Fig. 12.5 Two-state model for a base load unit

represents failure rate. Parameter μ represents the repair rate. Figure 12.5 shows two-state model for a base load unit. A power system is usually composed of a set of statistically independent components. In generating capacity reliability evaluation, these components are generating units that are described by two- or multi-state discrete capacity models.

The frequency of encountering certain state is the probability of being in the state multiplied by the rate of departure from the state. In the case of two-state model, it is also equal to the probability of not being in the state multiplied by the rate of entry.

There are three basic steps of frequency and duration method:

- Develop a suitable generation model from the parameters of the individual generating units.
- Develop a suitable load model from the given data over a defined period.
- Combine these two models to obtain the probabilistic model of the system capacity reserve or adequacy.

12.6.1 The Generation Model

There are two different approaches of developing the generation model: fundamental approach, which is not practical for large system analysis and recursive approach. If each unit can exist in two states, then there are 2^x states where x is number of units. Four-state diagram has 16 states and is already very complex. For multi-state unit, a recursive technique is rather used, which uses simple algorithms and can be practically used with a computer.

These algorithms give probability and frequency of having a given level of capacity forced out of service and of the complementary level of capacity in service. For practical use, it is better to combine the results to get cumulative probabilities and frequencies rather than the values corresponding to a specific capacity level. At any given capacity level, the cumulative values give the probability and frequency of having that capacity forced out of service. The mean of cumulative value here means that includes a specific value of power out of service or higher value of more power out of service.

One can add derated states after the two-state models have been derived with adding multi-state unit models. Figure 12.6 shows simple system with three states. The transitions between states are possible with the transition rate parameters μ and λ identified with the respective indexes.

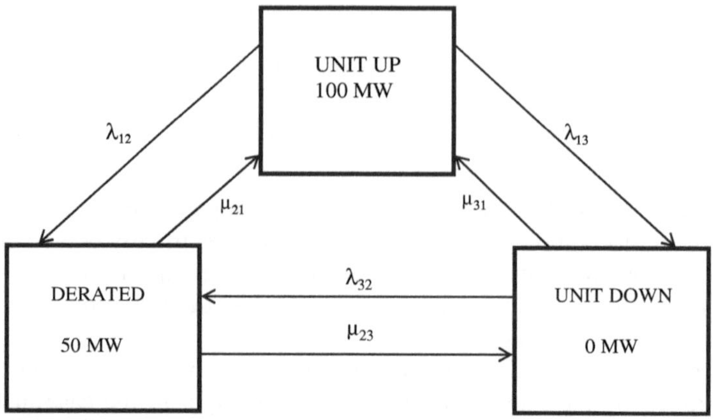

Fig. 12.6 Simple system with three states

12.6.2 System Risk Indices

Capacity models can be combined with the load models to get risk indices. Individual state load model or cumulative state load model can be used.

12.6.2.1 Individual State Model

Frequency and duration method is normally done on a period basis as the assumed constant low level and random peak load sequence do not usually apply over a long period of time. The behavior of the total system load can be expressed by a sequence of discrete load levels defined over the desired period of analysis. Figure 12.7 shows random period load model, where parameter L_i represents load level where its index i goes from the lowest to the peak load. The combination of discrete levels of available capacity and discrete levels of system demand or load creates a set of discrete capacity margins m_k. A margin is defined as the difference between the available capacity and the system load. A negative margin, therefore, represents a state in which the system load exceeds the available capacity and represents a system failure condition. The transition from one margin state to another can be made by change in load or a change in capacity. Basically, the frequency and duration method combines, through Markov techniques, capacity and load states to obtain the reserve or margin states.

The probability of the margin state is the product of the capacity state and load state probabilities:

$$p_k = p_n p_i \tag{12.10}$$

where p_n is the capacity state probability, and p_i is the load state probability.

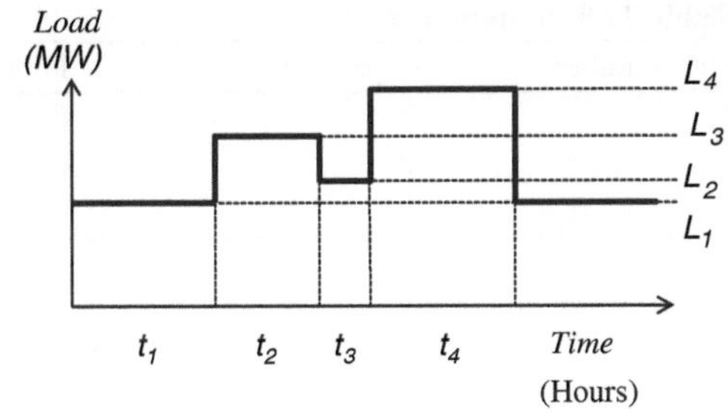

Fig. 12.7 Random period load model

The frequency of encountering the capacity margin state f_k represents the product of the steady state probability of the capacity margin state and the sum of the rates of departure from the state.

$$f_k = p_k(\lambda_{+k} + \lambda_{-k}) \tag{12.11}$$

where p_k is the probability of the margin state, λ indicates the rates of departure of the state, λ_{+k} is the transition rate to higher available capacity level, and λ_{-k} is the transition rate to lower available capacity level.

Simple examples of the method are shown in the next section. Several examples can be found in related references [12].

12.6.3 The Generation Model: Numerical Examples

12.6.3.1 Fundamental Approach

A simple numerical example is described in Table 12.9. The probability of being in a working state is 0.96 and the probability of being out of service is 0.04.

If each unit can exist in two states, then there are 2^x states where x is number of units. Figure 12.8 represents all possible states for a three unit state space diagram. Figure 12.8 shows the transition modes from one state to another [7].

The rate of departure from one state is the sum of individual rates of departure to other states. For example, for state 5, the rate of departure from state 5 is the sum of rates of departures: $\mu_2 + \mu_1 + \lambda_3$.

The state probability and state frequency is calculated in Table 12.10. The frequency of encountering certain state is the probability of being in the state multiplied by the rate of departure from the state. These rates are given in Table 12.9 and then used in the last column of Table 12.10.

Table 12.10 contains a number of identical capacity states that can be combined by summing obtained information (in our example, we can sum data from the states 2 and 3, 4 and 5, and 6 and 7). The results are shown in Table 12.11.

Table 12.11 gives the probability and the frequency of having a given level of capacity forced out of service and of the complementary level of capacity

Table 12.9 System data

Unit number	Capacity (MW)	Failure rate (day^{-1})	Repair rate (day^{-1})
1	20	0.02	0.48
2	20	0.02	0.48
3	40	0.02	0.48

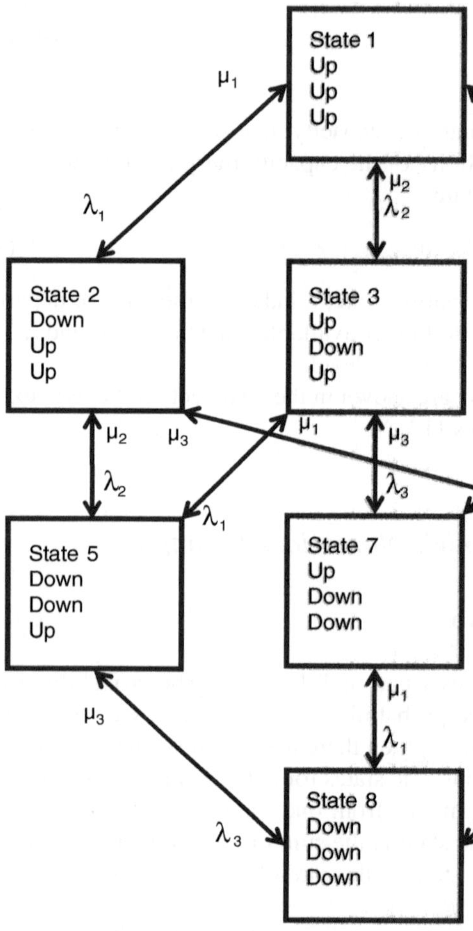

Fig. 12.8 Three-unit state space diagram

in service. But for practical use, it is better to have cumulative probabilities and frequencies rather than values corresponding to a specific capacity level.

Cumulative probability means that we are interested in a particular bad or worse situation.

Table 12.10 Generation model

State number	Capacity out C_i (MW)	State probability p_i	State frequency f (day^{-1})
1	0	0.96 × 0.96 × 0.96 = 0.884736	0.8847360 × (0.02 + 0.02 + 0.02) = 0.0530842
2	20	0.04 × 0.96 × 0.96 = 0.036864	0.0368640 × (0.48 + 0.02 + 0.02) = 0.0191693
3	20	0.96 × 0.04 × 0.96 = 0.036864	0.0368640 × (0.02 + 0.48 + 0.02) = 0.0191693
4	40	0.96 × 0.96 × 0.04 = 0.036864	0.0368640 × (0.02 + 0.02 + 0.48) = 0.0191693
5	40	0.04 × 0.04 × 0.96 = 0.001536	0.0015360 × (0.48 + 0.48 + 0.02) = 0.0015053
6	60	0.04 × 0.96 × 0.04 = 0.001536	0.0015360 × (0.48 + 0.02 + 0.48) = 0.0015053
7	60	0.96 × 0.04 × 0.04 = 0.001536	0.0015360 × (0.02 + 0.48 + 0.48) = 0.0015053
8	80	0.04 × 0.04 × 0.04 = 0.000064	0.0000640 × (0.48 + 0.48 + 0.48) = 0.00009216
		1.0	

Table 12.11 Reduced generation model

State number	Capacity out (MW)	Capacity in (MW)	Probability p_k	Frequency (occ day^{-1}, f_k)
1	0	80	0.8847360	0.0530842
2	20	60	0.036864 + 0.036864 = 0.073728	0.01916928 + 0.01916928 = 0.0383386
3	40	40	0.036864 + 0.001536 = 0.038400	0.01916928 + 0.00150528 = 0.0206746
4	60	20	0.001536 + 0.001536 = 0.003072	0.00150528 + 0.00150528 = 0.0030106
5	80	0	0.0000640	0.00009216

The individual state probabilities and frequencies can be combined to form the cumulative state values using the following equations:

$$P_{n-1} = p_k + P_n \tag{12.12}$$

$$F_{n-1} = F_n + p_k \lambda_{+k} - p_k \lambda_{-k} \tag{12.13}$$

Starting from the last, fifth state, the cumulative and individual values are the same. Capacity forced out is 80 MW (in theory, 80 MW or more, because we calculate cumulative values).

$$\begin{aligned} &C_5 \geq 80\,\text{MW} \\ &P_5 = p_5 = 0.000064 \\ &F_5 = f_5 = 0.00009216 \ \text{day}^{-1} \end{aligned}$$

State 4

$$\begin{aligned} &C_4 \geq 60\ \text{MW} \\ &P_4 = p_4 + P_5 = 0.003072 + 0.000064 = 0.003136 \\ &F_4 = F_5 + p_4\lambda_{+4} - p_4\lambda_{-4} = 0.00009216 + 0.00294912 - 0.00006144 \\ &\quad = 0.0029799\ \text{day}^{-1} \end{aligned}$$

Parameters $p_4\lambda_{+4}$ and $p_4\lambda_{-4}$ are calculated from p_x, which is one state probability that was calculated in Table 12.10, and transition rates $\lambda+$ and $\lambda-$, which can be found in Table 12.9.

Two 60-MW states (states 6 and 7) exist. The rate of departure of the state 6 is 0.48 + 0.48 + 0.02 as written in last column of Table 12.10. The transition value to a higher available capacity level is 0.48 + 0.48 and to lower level 0.02 (Fig. 12.8 shows the states). The rate of departure of the state 7 is 0.02 + 0.48 + 0.48. So:

$$p_4\lambda_{+4} = 0.001536 \times (0.48 + 0.48) + 0.001536 \times (0.48 + 0.48) = 0.00294912$$
$$\text{(State } 6 + \text{State } 7)$$
$$p_4\lambda_{-4} = 0.001536 \times (0.02) + 0.001536 \times (0.02) = 0.00006144$$
$$\text{(State } 6 + \text{State } 7)$$

State 3

$$C_3 \geq 40\,\text{MW}$$
$$P_3 = p_3 + P_4 = 0.0384 + 0.003136 = 0.041536$$
$$\begin{aligned} F_3 = F_4 + p_3\lambda_{+3} - p_3\lambda_{-3} &= 0.00297984 + 0.036865 \times (0.48) \\ &+ 0.001536 \times (0.48 + 0.48) - 0.036865 \times (0.02 + 0.02) \\ &- 0.001536 \times (0.02) \\ &= 0.0206438 \;\; \text{day}^{-1} \end{aligned}$$

State 2

$$C_2 \geq 20\,\text{MW}$$
$$P_2 = p_2 + P_3 = 0.073728 + 0.041536 = 0.115264$$
$$\begin{aligned} F_2 = F_3 + p_2\lambda_{+2} - p_2\lambda_{-2} &= 0.0206438 + 0.036864 \times (0.48 \times 2) \\ &- 0.036864 \times (0.02 \times 4) \\ &= 0.0530841 \;\; \text{day}^{-1} \end{aligned}$$

State 1

$$C_1 \geq 0\,\text{MW}$$
$$P_1 = p_1 + P_2 = 0.884736 + 0.115264 = 1.0000000$$
$$\begin{aligned} F_1 = F_2 + p_1\lambda_{+1} - p_1\lambda_{-1} &= 0.05308412 + 0.884736 \times (0.48 \times 0) \\ &- 0.884736 \times (0.02 \times 3) \\ &= 0 \;\; \text{day}^{-1} \end{aligned}$$

This approach is known as loss of capacity method. Results can be used directly as an indication of appropriate system generating capacity (Table 12.12).

Table 12.12 Generation system model

State number	Capacity out (MW)	Capacity in (MW)	Cumulative probability P_k	Cumulative frequency (occ day^{-1}, F_k)
1	0	80	1.0	0
2	20	60	0.115264	0.0530842
3	40	40	0.041536	0.0206438
4	60	20	0.003136	0.0029799
5	80	0	0.000064	0.0000922

12.6.3.2 Recursive Algorithm

The fundamental approach is not useful for larger systems. Calculating becomes too complex. More practical approach for large-system analysis is recursive technique. The algorithm can be easily computer processed. The technique can be used for two-state or multi-state unit and provides a fast technique for building capacity models (adding new units).

The method of calculating cumulative values ($P(X)$ and $F(X)$) is presented. The derated states can be added. The procedure is shown in reference [7].

Two-State (no derated states)

The recursive expression for a state of exactly X MW on forced outage after a unit of C MW and forced outage rate U is added are shown in equations below.

$$p(X) = p'(X)(1 - U) + p'(X - C)U \tag{12.14}$$

$$\lambda_+(X) = \frac{p'(X)(1 - U)\lambda(X) + p'(X - C)U(\lambda_+'(X - C) + \mu)}{p(X)} \tag{12.15}$$

$$\lambda_-(X) = \frac{p'(X)(1 - U)(\lambda_-'(X) + X)p'(X - C)(\lambda_-'(X - C))}{p(X)} \tag{12.16}$$

The $p(x)$, $\lambda_+(X)$, and $\lambda_-(X)$ parameters are the individual state probability and the upward and downward capacity departure rates, respectively, after the unit is added.

If X is less than C:

$$\begin{aligned} p'(X - C) &= 0 \\ \lambda_+'(X - C) &= 0 \\ \lambda_-'(X - C) &= 0 \end{aligned} \tag{12.17}$$

The numerical example includes the same system as in previous section and the expected results are the same. The system is given in Table 12.9.

Table 12.13 First unit added

State number	Capacity out (*X* MW)	Probability $p(x)$	$\lambda_+(x)$ (occ day^{-1})	$\lambda_-(x)$ (occ day^{-1})
1	0	0.96	0	0.02
2	20	0.04	0.48	0

Table 12.14 Calculation of $p(X)$ when second unit is added

Column 1: capacity out X (MW)	Column 2: $P'(1-U)$	Column 3: $P'(X-C)U$	Column 4: Column 4 = Column 2 + column 3
0	0.96 × 0.96	0 × 0.04	0.9216
20	0.04 × 0.96	0.96 × 0.04	0.0768
40	0 × 0.96	0.04 × 0.04	0.0016

Column 2: Expression 0 × 0.96: $p(40)$ has not yet been defined, because of this $p(40) = 0$
Column 3: Expression 0 × 0.04: $X = 0$ MW, $C = 20$ MW $\rightarrow X < C \rightarrow p'(X - C) = 0$ (see Eq. 12.7); Expression 0.96 × 0.04: $X = 20$, $C = 20 \rightarrow X = C \rightarrow p'(20\text{–}20) \rightarrow p'(0) = 0.96$ (see the Table 12.13); and Expression 0.04 × 0.04: $X = 40$, $C = 20 \rightarrow p'(40\text{–}20) \rightarrow p'(20) =$ 0.04 (see the Table 12.13)

Table 12.15 Calculation of $\lambda_+(X)$ when second unit is added

Column 1: capacity out X (MW)	Column 5: Col. 2 × $(\lambda_+'(X))$	Column 6: Col. 3 × $(\lambda_+'(X-C)+\mu)$	Column 7: Col. 5 + Col. 6	Column 8: Col. 7/Col. 4 $\lambda_+(x)$ (occ day^{-1})
0	0.9216000 × 0	0 × (0 + 0.48)	0	0
20	0.0384000 × 0.48	0.0384000 × (0 + 0.48)	0.0368640	0.4800000
40	0 × 0	0.0016000 × (0.48 + 0.48)	0.0015360	0.9600000

Column 5: Expression 0.9216 × 0: $X = 0 \rightarrow \lambda + '(X) = \lambda + '(0) = 0$ (see Table 12.13); Expression 0 × 0 $X = 40 \rightarrow \lambda_+'(X) = \lambda_+'(40)$ has not yet been defined, because of this $\lambda_+'(40) = 0$ (see Table 12.13)
Column 6: First row: $\lambda_+'(X - C) = 0$ because $X < C$ (see Eq. 12.7). For $\lambda_+'(20\text{–}20)$ and $\lambda_+'(40\text{–}20)$, see Table 12.13

Step 1: First, we add the first unit (Table 12.13)
Step 2: Then, we add the second 20-MW unit (Table 12.14)

All necessary data for next step are obtained. The third unit (40 MW) can be added.

Step 3: Adding the third unit. Proceed in the same way as before
Step 4: The individual capacity state probabilities are given in column 4 (Table 12.14). They can be combined directly with the values in column 8 (Table 12.15) and column 12 (Table 12.16) to give the individual state frequencies. These values can also be used to give the cumulative state probabilities using the following equations:

$$P(X) = P(Y) + p(X) \tag{12.18}$$

$$F(X) = F(Y) + p(X)(\lambda_+(X) - \lambda_-(X)) \tag{12.19}$$

Table 12.16 Calculation of $\lambda_{-}(X)$ when second unit is added

Column 1: capacity out X (MW)	Column 9: Col. 2 × ($\lambda_{-}(X) + \lambda$)	Column 10: Col. 3 × ($\lambda_{-}(X - C)$)	Column 11: Col. 9 + Col. 10	Column 12: Col. 11/Col. 4 $\lambda_{-}(X)$ (occ/day)
0	0.9216000 × (0.02 + 0.02)	0 × 0	0.0368640	0.0400000
20	0.0384000 × (0 + 0.02)	0.0384000 × 0.02	0.0015360	0.0200000
40	0 × (0 + 0.02)	0.0016000 × 0	0	0

Explanation: Practically the same as in Tables 12.14 and 12.15 ($\lambda_{-}(40)$ is not yet defined)

Table 12.17 Calculation of $p(X)$ when third unit is added

Column 1: capacity out X (MW)	Column 2: $P'(X)(1 - U)$	Column 3: $P'(X - C)U$	Column 4: Column 4 = Column 2 + Column 3
0	0.9216 × 0.96	0 × 0.04	0.884736
20	0.0768 × 0.96	0 × 0.04	0.073728
40	0.0016 × 0.96	0.9216 × 0.04	0.038400
60	0	0.0768 × 0.04	0.003072
80	0	0.0016 × 0.04	0.000064

Table 12.18 Calculation of $\lambda_{+}(X)$ when third unit is added

Column 1: capacity out X (MW)	Column 5: Col. 2 × ($\lambda_{+}'(X)$)	Column 6: Col. 3 × ($\lambda_{+}'(X - C) + \mu$)	Column 7: Col. 5 + Col. 6	Column 8: Col. 7/ Col. 4 $\lambda_{+}(x)$ (occ day^{-1})
0	0.884736 × 0	0 × (0 + 0.48)	0	0
20	0.073728 × 0.48	0 × (0 + 0.48)	0.0035389	0.48
40	0.001536 × 0.96	0.036864 × (0 + 0.48)	0.0019169	0.4992
60	0.0	0.003072 × (0.48 + 0.48)	0.0029491	0.96
80	0.0	0.000064 × (0.96 + 0.48)	0.0000922	1.44

Table 12.19 Calculation of $\lambda_{-}(X)$ when third unit is added

Column 1: capacity out X (MW)	Column 9: Col. 2 × ($\lambda_{-}(X) + \lambda$)	Column 10: Col. 3 × ($\lambda_{-}(X - C)$)	Column 11: Col. 9 + Col. 10	Column 12: Col. 11/Col. 4 $\lambda_{-}(X)$ (occ day^{-1})
0	0.884736 × (0.04 + 0.02)	0 × 0	0.0530842	0.06
20	0.073728 × (0.02 + 0.02)	0 × 0	0.0029491	0.04
40	0.001536 × (0 + 0.02)	0.036864 × (0.04)	0.0015053	0.0392
60	0 × (0 + 0.02)	0.003072 × (0.02)	0.0000614	0.02
80	0 × (0 + 0.02)	0.000064 × 0	0	0

where Y denotes the capacity outage state just larger than X MW.

Explanation: In previous Tables 12.17, 12.18, and 12.19, the results for $p(X)$, $\lambda_{+}(X)$, and $\lambda_{-}(X)$ exist.

Table 12.20 Complete generation model

Capacity out X (MW)	Probability $p(X)$	$\lambda_+(X)$ (occ day^{-1})	$\lambda_-(X)$ (occ day^{-1})	Frequency (occ day^{-1}) f_x	Cumulative probability $P(X)$	Cumulative frequency (occ day^{-1}) $F(X)$
0	0.884736	0	0.06	0.0530842	1.0	0
20	0.073728	0.48	0.04	0.0383386	0.115264	0.0530842
40	0.0384	0.4992	0.0392	0.0206746	0.041536	0.0206439
60	0.003072	0.96	0.02	0.0030106	0.003136	0.0029799
80	0.000064	1.44	0	0.00009216	0.000064	0.00009216

$f_x = p(X)(\lambda_+(X) + \lambda_-(X))$: The frequency of encountering certain state is the probability of being in the state multiplied by the rate of departure from the state.

$P(X)$: Use Eq. 12.18 and start with P(80) that is the same as $p(X)$.

$P(60) = 0.000064 + 0.003072 = 0.003136$

$P(40) = 0.003136 + 0.0384 = 0.041536$

$F(X)$: Use Eq. 12.19 and start with F(80), which is the same as f_x. The same principle as for $P(X)$.

Be careful: In Eq. 12.19, λ_+ and λ_- are subtracted, unlike when calculating f_x.

We can see that results with two different techniques are the same (Tables 12.12 and 12.20). For large systems, the recursive algorithms should be used.

References

1. Calabrese G (1947) Generating reserve capacity determined by the probability method. AIEE Trans 66:1439–1450
2. Adequacy assessment of generating units in Slovenia for the Period 2005–2008 (2005) (in Slovenian). ELES
3. Brezovec B, Matko V, Podberšič M (2009) Concept of reliability design of modern informatic system for emergency call. FERI Maribor
4. Rietz R (2006) Costs of adequacy and reliability of electric power. IEEE
5. Vijayamohanan PN (2008) Loss of load probability of a power system. Munich
6. Stoll HG (1989) Least-cost electricity utility planning. Wiley, New York
7. Billinton R, Allan R (1996) Reliability evaluation of power systems. Plenum, New York
8. Leite da Silva AM, Melo ACG, Cunha SHF (1991) Frequency and duration method for reliability evaluation of large-scale hydrothermal generating systems. In: Proceedings of Inst Electr Eng C138(1)
9. Melo ACG, Pereira MVF, Leite da Silva AM (1992) Frequency and duration calculations in composite generation and transmission reliability evaluation. IEEE Trans Power Syst 7(2):469–476
10. Melo ACG, Pereira MVF, Leite da Silva AM (1993) A conditional probability approach to the calculation of frequency and duration indices in composite reliability evaluation. IEEE Trans Power Syst 8(3):1118–1125
11. Renuga P, Ramaraj N, Primrose A (2006) Frequency and duration method for reliability evaluation of large scale power generation system by fast fourier transform technique. J Energy Environ 5

12. Billinton R, Allan RN (1988) Reliability assessment of large electric power systems. Kluwer, Boston
13. Wang X, McDonald JR (1994) Modern power system planning. McGraw-Hill, New York
14. Clifford E (2009) Overview of AdCal model. EirGrid, Ireland
15. World Energy Council (2004) Performance of generating plant: new realities, new needs. A report of the World Energy Council
16. Allan RN, Billinton R (1992) Probabilistic methods applied to electric power systems: are they worth it? Power Eng J 6(3):121–129B
17. IEEE Std 1366 (2003) Guide for electric power distribution reliability indices. IEEE
18. Elmakias D (2008) New computational methods in power system reliability. Springer, Berlin, Heidelberg
19. Anders GJ (1990) Probability concepts in electric power systems. Wiley, New York

Chapter 13
Reliability and Performance Indicators of Power Plants

It is remarkable that a science which began with the consideration of games of chance should have become the most important object of human knowledge

Pierre Simon Laplace

13.1 Introduction

The reliability of power plants is one of the parameters of the reliability of power systems. The power system includes a variety of power plants so each of the plants is presented by its reliability indicators. The need for several indicators for one plant arises from the fact that the plants under consideration are fairly complex facilities and they are a part of a very complex system, where only one indicator may not be sufficient.

Actually, the contents about the reliability indicators have been expanded, so the presented indicators exceed the definition of reliability itself and they give some information also about related questions. For example, we can have very reliable hydro power plant, but if there is no water, the availability of the plant would be very low and the capacity would be very low. The reliability itself would give the incomplete information. Similarly, it is so for wind power plant, if there is no wind, or for solar power plant if there is not enough sun.

The presented indicators collected for distinguished power plants are not all comparable to each other. The reason for their incomparability lays in a fact that different groups of professionals deal with each of the plants and they do not invest the efforts in some unification of the terminology and methodology.

For some power plant, three groups of operation indicators are defined: (i) technical indicators, (ii) environmental indicators, and (iii) sociological indicators; for some power plants, there are less groups, if any.

Operation indicators represent power plant in several aspects, including the technical safety, reliability, installation performance, waste generation, personnel safety, and the effects to environment. The indicators are intended for managing plant for improving performance, for setting the goals for improvement, and for providing the means to assess the overall plant performance [1–5].

The use of the indicators is not limited to the specific plant where they are defined in the following sections. The indicators can be used widely. The selection

M. Čepin, *Assessment of Power System Reliability*,
DOI: 10.1007/978-0-85729-688-7_13, © Springer-Verlag London Limited 2011

of indicators at each specific plant was selected based on the expected applications, which may vary.

13.2 Nuclear Power Plant

13.2.1 Purposes and Definitions of the Indicators

The main purpose of the indicators is to communicate the recent developments in the area of nuclear power plant safety.

13.2.2 Unit Capability Factor

The purpose of unit capability factor is to monitor progress in attaining high unit energy production reliability. This indicator reflects effectiveness of plant programs and practices in maximizing available electrical generation and provides an overall indication of how well plant is being operated and maintained.

Unit capability factor is defined as the ratio of the available energy generation over a given time period to the reference energy generation over the same time period expressed as a percentage.

Available energy generation is the energy that could have been produced under reference ambient conditions considering only limitations within control of plant management, i.e., plant equipment and personnel performance, and work control.

Reference energy generation is the energy that could be produced if the unit was operated continuously at full power under reference ambient conditions.

$$\mathrm{UCF} = \frac{(\mathrm{REG} - \mathrm{PEL} - \mathrm{UEL}) \times 100\%}{\mathrm{REG}} \tag{13.1}$$

where UCF is the Unit Capability Factor expressed as a percentage, REG is the Reference Energy Generation for the period, PEL is the total Planned Energy Losses for the period, and UEL is the total Unplanned Energy Losses for the period.

Operation below referenced unit capacity because of environmental limitations is not considered as lost energy generation and not used in calculations.

13.2.3 Unplanned Capability Loss Factor

The purpose of unplanned capability loss factor is to monitor progress in minimizing outage time and power reductions that result from unplanned equipment failures or other conditions. This indicator is defined as the ratio of the unplanned energy losses during a given period of time to the reference energy generation expressed as a percentage.

$$\mathrm{UCLF} = \frac{\mathrm{UEL} \times 100\%}{\mathrm{REG}} \tag{13.2}$$

where UCLF is the unplanned capability loss factor expressed as a percentage, and UEL is the total unplanned energy losses for the period.

13.2.4 Unplanned Automatic Scrams per 7,000 h Critical

This indicator is defined as the number of unplanned automatic scrams because of reactor protection that occur per 7,000 h of critical operation and is calculated as follows:

$$\text{UAS/7,000 h critical} = \frac{\text{Number of unplanned autoscrams} \times 7,000}{\text{Total number of hours critical}} \tag{13.3}$$

13.2.5 Thermal Performance

The purpose of thermal performance indicator is to monitor progress in achieving and maintaining efficient thermal operation. It provides an indication of success toward meeting the design capabilities.

This indicator is defined as the ratio of the design gross heat rate to the adjusted actual gross heat rate. If the adjusted actual gross heat rate values are routinely found to be lower than the design gross heat rate value, then new design heat rate value is calculated based on the lowest adjusted actual gross heat rate that has been measured during routine thermal performance monitoring activities. Thermal performance indicator is calculated as follows:

$$\text{Thermal performance} = \frac{\text{Design gross heat rate} \times 100\%}{\text{Adjusted actual heat rate}} \tag{13.4}$$

13.2.6 Collective Radiation Exposure

The purpose of this indicator is to monitor efforts to minimize total radiation exposure of plant personnel and to measure the effectiveness of radiological protection programs. Collective radiation exposure is total external whole-body dose received by all people onsite, i.e., personnel including contractors and visitors during a time period, as measured by the primary dosimeter, thermo luminescent, or a film badge.

Exposure measured by direct reading dosimeters should be included only for those periods or situations when more accurate data are not available from thermo luminescent or film badges.

13.2.7 Volume of Low-Level Solid Radioactive Waste

The purpose of the low-level solid radioactive waste indicator is to monitor progress toward reducing the volume of low-level waste production, which will decrease storage, transportation, final disposal needs, and improve public perception of the environmental impact of nuclear power.

This indicator is defined as the volume of low-level solid radioactive waste that has been processed and is in final form ready for burial during given period. The volume of radioactive waste that has not completed processing and is not yet in final form is not included. Low-level solid radioactive waste consists of dry active waste, sludge, resins, and evaporator bottoms generated as a result of nuclear power plant operation and maintenance. Low-level radioactive waste refers to all radioactive waste that is not spent fuel or a by-product of spent fuel processing.

13.2.8 Industrial Safety Accident Rate

The purpose of industrial safety accident rate is to monitor progress in improving safety performance for utility personnel permanently assigned to the station. Industrial safety accident rate is the number of accidents resulting in one or more days out of work excluding the day of the accident, or one or more days of restricted work excluding the day of the accident, and work-related fatalities per 200,000 man-hours worked. Contractor personnel are not included for this indicator. Industrial safety accident rate is calculated according to equation:

$$\mathrm{ASAR} = \frac{(\text{Number of lost time accidents} + \text{number of fatalities}) \times 200{,}000}{\text{Number of station man hours worked}} \tag{13.5}$$

13.2.9 Safety System Performance

The purpose of this indicator is to monitor the readiness of the important safety systems to respond to abnormal events or accidents. This indicator monitors the effectiveness of operation and maintenance practices in reducing the unavailability of safety system components. Safety system performance indicator is calculated separately for each of the following three pressurized reactor systems:

- High-pressure safety injection
- Auxiliary feedwater system
- Emergency alternating current power supply system

It is defined for each safety system as the sum of the hours system/component was unavailable and estimated unavailable hours during a time period divided by product of hours system is required during that time period and numbers of trains in the system. This definition is further explained as follows:

- Component unavailability hours is the fraction of time that a component is unable to perform its intended function when it is required to be available for service.
- Estimated unavailability hours are the average hours of a component being in a failed state before discovery of a failure. It could be zero if the exact time of failure is known.

Component is the equipment for which the unavailable hours are recorded. A component is included in the safety system performance indicator when unavailability of the component can degrade the full capacity or redundancy of the system. Safety system performance is calculated as follows:

$$\mathrm{SSP} = \frac{\text{Known unavailability hours} + \text{estimated unavailability hours}}{\text{Hours system required} \times \text{number of trains}} \tag{13.6}$$

13.2.10 Fuel Reliability

The purpose of the fuel reliability indicator is to monitor progress in achieving and maintaining high fuel integrity. This indicator is defined as the steady-state primary coolant activity of Iodine I–131, which is an important radioisotope of iodine, corrected for the tramp uranium contribution and normalized to a common purification rate and linear heat generation rate. The activity of Iodine is expressed in units of $\mu\mathrm{Ci\ g}^{-1}$. Fuel reliability indicator is calculated as follows:

$$\mathrm{FRI} = \left[(A_{131})_N - k \times (A_{134})_N\right] \times \sqrt{\left(\frac{L_N}{\mathrm{LHGR}} \times \frac{100}{P_0}\right)^3} \tag{13.7}$$

where FRI is the fuel reliability indicator, $(A_{131})_N$ is the average steady-state activity of I-131 in the coolant normalized to a common purification rate and expressed in μCi/g, $(A_{134})_N$ is the average steady state activity of I-134 in the coolant normalized to a common purification rate and expressed in μCi/g, K is the tramp correction coefficient (a constant with a value of 0.0318) and this coefficient is based on a tramp material composition of 30% uranium and 70% plutonium,

L_N is the linear generation rate at 100% power for the unit, LHGR is the linear heat generation rate, and P_0 is the average reactor power expressed in percentage at the time activities are measured.

13.2.11 Chemistry Index

The purpose of this indicator is to evaluate and trend progress in improving chemistry control. The calculation by itself is based on concentration of important impurities in plant systems. These are monitored impurities for pressurized water reactor with recirculation steam generators:

- Steam generator blowdown cation conductivity
- Steam generator blowdown sodium
- Condensate pump discharge dissolved oxygen

The chemistry index combines information about several important chemistry parameters into a single indicator that can be used as a management tool to provide an overview of the effectiveness of the chemistry program. Chemistry index is calculated as follows:

$$\text{Chemistry index} = \frac{\frac{K_a}{0.8} + \frac{\text{Na}}{20} + \frac{\text{O}_2}{10}}{3} \tag{13.8}$$

where K_a is the average blowdown cation conductivity (μS/sm at 25°C), Na is the average blowdown sodium concentration (ppb), and O_2 is the average condensate pump discharge dissolved oxygen concentration (ppb).

13.2.12 Time Availability Factor

Time availability factor is defined as ratio of the unit available hours in a given period to the total number of hours in the same period, expressed as a percentage. Time period calculated for the whole year 1997 is 8,784 h.

Unit available hours are the total number of hours in a given period while the unit is operating online, or is capable of such operation. It is calculated as a sum of total generator online hours and total unit reserve shutdown hours for a given period.

Unit reserve shutdown hours are the total number of hours in a given period during the unit is removed from online operation for economic or other similar reasons, when operation could have been continued.

Time availability factor is calculated according to equation:

$$\text{TAF} = \frac{(\text{GEN} + \text{URS}) \times 100\%}{\text{HRS}} \tag{13.9}$$

where TAF is the time availability factor expressed as a percentage, GEN is the total generator online hours in the period, URS is the total unit reserve shutdown hours in the period, and HRS is the total number of hours in the period.

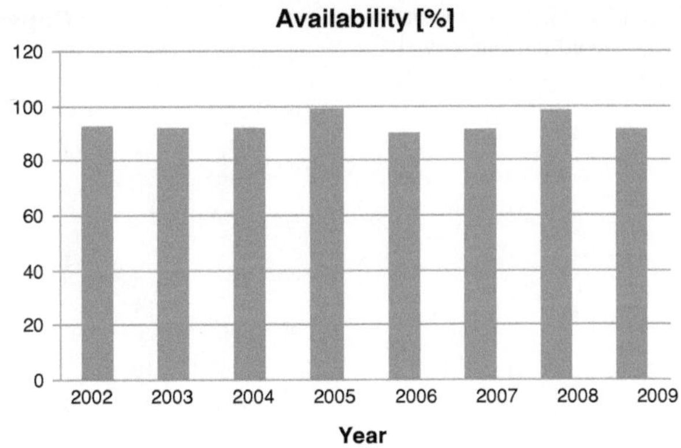

Fig. 13.1 Availability for the selected nuclear power plant

Figure 13.1 shows availability for the selected nuclear power plant in recent years [6–13].

13.2.13 Monthly Time Availability Factor

The same definition applies as for the time availability factor, but it is calculated for a period of each month in the year separately.

13.2.14 Capacity Factor (load factor)

The capacity factor is the ratio of the energy produced during the given period to the energy that could have been produced at maximum capacity under continuous operation during the whole of that period. Time period calculated for the whole year 1997 is 8,784 h. Capacity factor calculation:

$$\mathrm{CAF} = \frac{\mathrm{NEP} \times 100\%}{\mathrm{MDC} \times \mathrm{HRS}} \tag{13.10}$$

where CAF is the capacity factor expressed as a percentage, NEP is the total net energy production in the period, and MDC is the maximum dependable capacity.

Figure 13.2 shows the capacity factor for the selected nuclear power plant in recent years [6–13].

13.2.15 Monthly Capacity Factor (monthly load factor)

The same definition applies as for the capacity factor, but it is calculated for each month separately.

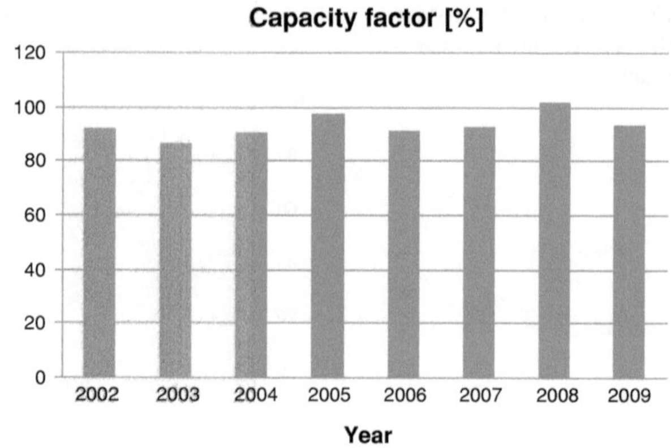

Fig. 13.2 Capacity factor for the selected nuclear power plant

13.2.16 Net Electrical Energy Production

Net electrical energy production gives a total amount of net electrical energy produced during a year and delivered to 400 kV network.

13.2.17 Monthly Net Electrical Energy Production

Monthly net electrical energy production gives a total amount of net electrical energy produced during a month and delivered to 400 kV network.

13.2.18 Number of Unplanned Automatic Scrams While Critical

The indicator is defined as the number of unplanned automatic scrams (reactor protection system logic actuation) that occur while the reactor is critical. Manual reactor trip or a trip that follows a manual turbine trip as a result of operator's intention to protect equipment or mitigate the consequences of a transient are not counted, because operator actions to protect equipment should not be discouraged.

13.2.19 Number of Unplanned Safety Injection Actuation

This indicator is defined as the number of unplanned safety injection actuations that result from reaching actuation setpoint or from an inadvertent safety injection signal.

13.2.20 Duration of Annual Outage

Duration of annual outage stands for the total duration of annual outage in hours.

13.3 Thermal Generating Power Plant

13.3.1 Forced Outage Rate

The forced outage rate (FOR) is the basic generating unit parameter used in static capacity evaluation and it represents the probability of finding the unit on forced outage at some distant time in the future. It can be better defined as unit unavailability, as it is not expressed in unit of a number per time period as the rates usually are.

The forced outage rate is defined according to the equation below.

$$\text{FOR} = \frac{\text{FOH}}{\text{FOH + SH}} = \frac{\lambda}{\lambda + \mu} = \frac{r}{r + m} = \frac{r}{T} = \frac{f}{\mu} = \frac{\sum[\text{down\_time}]}{\sum[\text{down\_time}] + \sum[\text{up\_time}]} \quad (13.11)$$

where FOH is the full forced outage hours = unavailability, SH is the service hours, λ is the expected failure rate, μ is the expected repair rate, m is the mean time to failure: MTTF = $1/\lambda$, r is the mean time to repair: MTTR = $1/\mu$, $m + r$ is the mean time between failures: MTBF = $1/f$, f is the cycle frequency: $f = 1/T$, and T is the cycle time: $T = 1/f = m + r$.

13.3.2 Equivalent Forced Outage Rate

Equivalent forced outage rate represents the probability that a generating unit will not meet its required generation demanded by dispatch [14]. EFOR is defined according to the equation below [3].

$$\text{EFOR} = \frac{(f_d \times \text{FOH}) + (\text{EFDH} - \text{EFDHRS}) \times 100\%}{\text{SH} + (f_d \times \text{FOH})} \quad (13.12)$$

FOH	Full forced outage hours
EFDH	Equivalent forced derated hours
EFDHRS	Equivalent forced derated hours during reserve shutdown
SH	Service hours
f_d	Discount factor for FOH

$$f_d = (1/r + 1/T)/(1/r + 1/T + 1/D) \quad (13.13)$$

r	Average forced outage duration
T	Average reserve shutdown time
D	Average demand time
$T + D$	Available hours/number of starts

13.3.3 Unit Capability Factor

Unit capability factor is the percentage of maximum energy generation that a plant can produce to electrical grid. It is usually limited by factors within plant control management.

A high value of unit capability factor indicates effective plant program, minimizing unplanned energy losses, maximizing available generation, and optimizing planned outages.

13.3.4 Unplanned Capability Loss Factor

Unplanned capability loss factor is percentage of maximum energy generation that a plant is not capable to produce to electrical grid because of unplanned energy losses.

A low value of unplanned capability loss factor indicates that plant equipment is reliably operated and well maintained.

13.3.5 Unplanned Automatic Grid Separations per 7,000 Operating Hours

This indicator shows how often a unit is separated from the external grid in unplanned and automatic manner. It is given as a rate per 7,000 operating hours.

13.3.6 Successful Start-Up Rate

This indicator expresses the level of unit successful presence in electrical grid at the moment requested.

13.3.7 Industrial Safety Accident Rate

Industrial safety performance is monitored by the number of accidents that result a day away from work per 1,00,0000 man-hours worked.

13.3.8 Commercial Availability

Commercial availability stands for a conception that includes the influence of the price/cost gap magnitude so that it could serve as more accurate indicator of the plant impact on the company profit. It should reflect the relationship between supply and demand with the value of MWh increasing because of an increase in demand or decrease in supply. Commercial availability is a very popular indicator despite the fact that there is no specific definition of the term. There are few methods that evaluate commercial availability on historical basis [14]:

- Method 1 (goal commercial availability) compares the ratio of the actual income to potential income using the product of generation and market price:

$$\mathrm{CA} = \frac{\text{Actual generation} \times \text{market price}}{\text{Requested generation} \times \text{market price}} \tag{13.14}$$

- Method 2 (traditional availability) is based on the ratio of hours during which the unit was available to the total hours that the unit could have been operating at profit:

$$\mathrm{CA} = \frac{\text{Hours available}}{\text{Period hours}} \tag{13.15}$$

- Method 3 is based on the ratio between accumulated sum of the difference between market price and generating costs multiplied with maximum available capacity of the unit to accumulated sum of the difference between market price and generating costs multiplied with maximum available capacity in each hour:

$$\mathrm{CA} = \frac{\sum (\text{market price} - \text{generation cost}) \times \text{MW available}}{\sum (\text{market price} - \text{generation cost}) \times \text{MAX capacity}} \tag{13.16}$$

- Method 4 (actual commercial availability) is based on ratio between actual margin (sum of the generation during selected period) and potential margin (installed capacity in the place during time period):

$$\mathrm{CA} = \frac{\text{Actual margin}}{\text{Potential margin}} \tag{13.17}$$

- Method 5 is based on ratio between actual generation multiplied with the difference between market price and unit cost and planned availability multiplied with the difference between market price and unit cost:

$$\mathrm{CA} = \frac{\text{Actual generation} \times (\text{market price} - \text{unit cost})}{\text{Planned availablity} \times (\text{market price} - \text{unit })} \tag{13.18}$$

Commercial availability can also be forecasted and is sometimes referred to as financial availability. To evaluate the future commercial availability for a specific generating unit, it is necessary to develop a forecast of market prices on an hourly basis. A market simulation model is used to project prices and revenue. The model has to include transmission congestion, regional transmission organization impacts, and an hour-by-hour dispatch. Output of the market forecast determines the highest value of commercial availability opportunities.

Commercial availability requires a different approach in applying data and new tools in day-to-day and performance assessment decisions according to a wide range of impacts:

- *Benchmarking—selection*: Indentifying plants whose design and operational characteristics are similar to the unit in question.
- *Benchmarking—comparisons*: Commercial availability is usually dependent on the market price per hour that can be matched to unit's availability. By combining the goal conditional probability and a unit's unique economics, it is possible to calculate a goal for commercial availability without creating any new data collection processes.
- *Maximizing commercial availability* is focused on generating when required by market and when the income and profit are the highest. Units are maintained and manned to meet market need. When unit is not required by the market and is technically unavailable, it has no effect on commercial availability.
- *Design*: New plant design can be affected because the goal is not to maximize availability or reliability, but to maximize profit. Therefore, dependency can be reduced.

13.3.9 Environmental Indicators

The environmental indicators include quantities of emissions of pollutants, such as SO_2, NO_2, or ashes, for example. They are expressed in absolute or more often in relative terms considering the observed time interval or considering the energy produced.

13.4 Geothermal Power Plant

The main geothermal power facility performance indicators include three main indicators that describe the technical geothermal power facility performance [15]:

- Capacity factor
- Load factor
- Availability factor

Those three indicators are dimensionless and can be expressed in percentages. Other indicators include safety accident rate, production loss control, environmental indicator, etc.

13.4.1 Capacity Factor

Capacity factor is needed to describe the technical performance of the plant. It is calculated by ratio between the total electrical energy generated and the installed capacity multiplied with the time period.

$$\text{CF} = \frac{\text{Total energy generated}}{\text{Installed capacity} \times \text{period}} \tag{13.19}$$

13.4.2 Load Factor

Load factor is defined as ratio between the total electrical energy generated and the maximum load multiplied with the time period.

$$\text{LF} = \frac{\text{Total energy generated}}{\text{Maximum load} \times \text{period}} \tag{13.20}$$

When the capacity factor is the same as load factor, then the installed capacity matches the field conditions and the market conditions. When the capacity factor is much lower than load factor, the installed capacity is too large.

13.4.3 Availability Factor

$$\text{AF} = \frac{\text{Total hours of operation during period}}{\text{Period time}} \tag{13.21}$$

Two separate availability factors are defined including the time lost during the planned outage.

$$\text{Steam AF} = \frac{\text{Average steam flow during period}}{\text{Steam supply}} \tag{13.22}$$

$$\text{Steam AF} = \frac{\text{Steam supply steam production shortfall}}{\text{Steam supply}} \tag{13.23}$$

The specific operational cost is an indicator of the operational cost per kWh at the generator transformer supply voltage terminals.

$$\text{Specified operational cost} = \frac{\text{Cost for period}}{\text{Energy generated for period}} \times 1{,}000 \tag{13.24}$$

13.4.4 Safety Accident Rate

Safety accident rate is defined as a number of accidents for all personnel permanently assigned to geothermal power facility per 1,000,000 man-hours worked.

13.4.5 Production Loss Control

Production loss control is the number of forced outages per period per generating facility.

13.4.6 Environmental Indicator

The environmental indicator for geothermal energy is defined as the quantity of H_2S emissions (Q_{H_2S}) usually expressed in grams per kilowatt hour (g/kWh).

13.5 Hydroelectric Power Plant

For hydroelectric plants, the indicators such as described previously can be used for the complete plant or for each generator and its respective equipment separately or both: availability, capability factor, safety accident rate, and forced outage rate [16–20].

13.6 Biomass Power Plant

The indicators such as described previously can be used: availability, capability factor, load factor, safety accident rate, and forced outage rate.

Environmental indicators include quantities of ash emitted (Q_{ash}) expressed in grams per kilowatt hour (g/kWh) and quantities of CH_4 emitted from landfills or animal manure expressed in grams per kilowatt hour (g/kWh).

13.7 Solar Power Plant

The performance of solar power plant depends on weather, especially sun conditions. Both the site and the region, where the plant is installed, are very important.

The most important performance indicators are difficult to measure as they depend on the exact position of the sun. Therefore, specific measurement instruments are required. The performance indicators include reference yield, array yield, final yield, and performance ratio.

The electrical properties of photovoltaic devices are given at standard test conditions (STC): cell temperature at 25°C, solar irradiance at 1,000 W/m^2, and solar spectrum at air mass 1.5. The power maximum under standard test conditions is named peak power (W_p).

13.7.1 Reference Yield

Reference yield (Y_R) is defined as solar irradiation on the tilted plane normalized to the solar irradiance under standard test conditions in a day or month or year.

$$Y_R = \frac{E_{S,A}}{G_{\text{STC}}} \tag{13.25}$$

where $E_{S,A}$ is the in-plane irradiation (Wh/m^2), and G_{STC} is the reference in-plane irradiance = 1 kW/m^2.

It is expressed in hours or in kWh/kW$_p$ in a day or month or year and can be considered as the number of hours during which the solar radiation would be at reference irradiance level in a day or month or year [14].

13.7.2 Array Yield

Array yield (Y_A) is the daily or monthly or yearly energy array output per kWp of installed array power:

$$Y_A = \frac{E_{A,d}}{P_0} \tag{13.26}$$

where $E_{A,d}$ is the energy array output (kWh/d), and P_O is the installed array power (kW$_p$).

It is expressed in kWh/(d kWp) and is considered as the number of hours of array operation per day at installed array power, which would give the same energy output as the recorded integral value for that day, month, or year.

13.7.3 Final Yield

Final yield (Y_f) is daily or monthly or yearly plant useful energy output per kWp of installed array power:

$$Y_f = \frac{E_{\text{USE}}}{P_0} \tag{13.27}$$

where E_{USE} is the plant useful energy output (kWh/d).

It is expressed in kWh/(d kWp) and it represents the number of hours of plant operation per day at installed array power, which would give the same energy output as the recorded integral value for that day, month, or year.

The losses can be calculated considering the losses before and passing the inverter, respectively, which are capture losses and system losses.

$$L_{\mathrm{C}} = Y_R - Y_A \tag{13.28}$$

$$L_{\mathrm{S}} = Y_A - Y_F \tag{13.29}$$

where L_{C} is the capture losses, and L_{S} is the system losses.

The term system losses may be misleading. It includes all losses that are not capture losses. For a grid connected photovoltaic installation, the system losses are mainly inverter losses.

13.7.4 Performance Ratio

Performance ratio indicates the overall effect of losses on the array rated output because of array temperature, incomplete use of the irradiation, and system component inefficiencies or failures.

$$\mathrm{PR} = \frac{Y_f}{Y_R} \tag{13.30}$$

13.7.5 Environmental Indicators

The environmental indicator for solar power plant is related to toxic materials in cells and batteries. It is defined as quantity of toxic materials (Q_{tox}) contained in the cells and batteries that will be recycled or disposed after their lifetime expressed in grams per peak power (g/Wp).

13.8 Wind Power Plant

Wind power plant performance largely depends on wind turbines and on wind conditions on the site. Therefore, along with technical performance indicators also wind regime must be considered.

Wind measurements should be carried out with rate of 1 Hz and documented using parameters such as:

- Mean wind speed
- Weibull distribution
- Directional distribution of wind energy
- Turbulence intensity

To make power output of different wind turbines model comparable, their performance is measured and documented. The curves are usually only related to standard conditions and not useful with precision to the various installations on site.

13.8.1 Electricity Production Indicators

- Total energy production is delivered to the grid during selected period, usually one year.
- Specific energy production per square meter depends on the site and is expressed in kilowatt hours per square meter (kWh/m^2).
- Equivalent full load hours are annual energy production in relation to the rated power of the turbine in hours.
- Capacity factor is the ratio of the total energy production during 1 year over the potential energy production.

13.8.2 Technical Availability Indicators

- Nominal period is the complete period covered by the report and is usually 1 year.
- Period of non-availability is the period during which the plant is not available for generation. Non-availability can be scheduled or unscheduled.
- Technical availability is the period of availability over the nominal period in percentage.
- Average technical non-availability divides the total period of non-availability by the number of considered turbines.

13.8.3 Additional Possible Indicators Related to Weather Variability Control

Accurate forecast of the wind and sun conditions is very important for an efficient and economical use of wind power plants. Therefore, we try to define indicators related to the forecast of the weather conditions.

For this reason, two indicators can be defined. One is related to the estimation of the electric energy production, whereas the other one to the prediction of the money value.

$F_d[\%]$ = Effective kWh (or money) produced during day d (week w, month m, and year y),

F_w, F_m, F_y = kWh (or money) forecasted 1 day (week, month, and year) before.

13.8.4 Environmental Indicators

There are four environmental indicators specific for wind power plant: (i) visual effect (landscape protection distance), (ii) noise from wind turbines, (iii) bird fatalities, and (iv) shadow casting [14].

References

1. Billinton R, Allan RN (1988) Reliability assessment of large electric power systems. Kluwer, Boston
2. Data Reporting Instructions (2004) Generating availability data system. NERC
3. World Association of Nuclear Operators (2008) 2007 Performance indicators. WANO
4. World Association of Nuclear Operators (2009) 2008 Performance indicators. WANO
5. Nuclear Power Plant Krško (1998) Performance indicators for year 1997. ESD Analysis and Licensing, NEK
6. Annual report 2009 on the radiation and nuclear safety in the Republic of Slovenia (2010) URSJV
7. Expanded annual report on the radiation and nuclear safety in the Republic of Slovenia for the year 2009 (2010) (in Slovenian). URSJV
8. Annual report 2008 on the radiation and nuclear safety in the Republic of Slovenia (2009) URSJV
9. Expanded annual report on the radiation and nuclear safety in the Republic of Slovenia for the year 2008 (2009) (in Slovenian). URSJV
10. Annual report 2007 on the radiation and nuclear safety in the Republic of Slovenia (2008) URSJV
11. Expanded annual report on the radiation and nuclear safety in the Republic of Slovenia for the year 2007 (2008) (in Slovenian). URSJV
12. Annual report 2006 on the radiation and nuclear safety in the Republic of Slovenia (2007). URSJV
13. Expanded annual report on the radiation and nuclear safety in the Republic of Slovenia for the year 2006 (2007) (in Slovenian). URSJV
14. World Energy Council (2004) Performance of generating plant: new realities, new needs: a report of the World Energy Council
15. International Geothermal Association for the World Energy Council (2001) Performance indicators for geothermal power plants
16. Chirikutsi J (2007) Plant performance measurement in hydro power plants. In: Proceedings of the 19th African hydro symposium
17. Chowdhury A, Koval DO (2000) Development of transmission system reliability, performance benchmarks. IEEE Trans Ind Appl 36(3):899–903
18. Allan RN, Billinton R (1992) Probabilistic methods applied to electric power systems: are they worth it? Power Eng J 6(3):121–129
19. IEEE Std 1366 (2003) Guide for electric power distribution reliability indices
20. Layton L (2004) Electric system reliability indices

Chapter 14
Distribution and Transmission System Reliability Measures

To conquer without risk is to triumph without glory
Pierre Corneille

14.1 Introduction

The power system reliability is one of the features of power system quality in addition to required voltage and constant frequency. The electric utility industry has developed several performance measures of reliability or reliability indices [1–24]. These reliability indices include measures of outage duration, frequency of outage, number or customers involved or their lost power or energy, and the response time. The Institute of Electrical and Electronic Engineers (IEEE) defines the generally accepted reliability indices in its standard number [1]. This standard lists several important definitions for reliability, including what are momentary interruptions, momentary interruption events, and sustained interruptions.

The standard distribution and transmission reliability indices and factors that affect their calculation are collected and presented. The indices are intended to apply to power distribution and transmission systems, substations, circuits, and defined regions.

14.2 Distribution Reliability Indices

The distribution and transmission reliability indices include system average interruption frequency index (SAIFI), transformer SAIFI, equivalent number of interruptions related to the installed capacity (NIEPI), customer interruption, system average interruption duration index (SAIDI), transformer SAIDI, equivalent interruption time related to the installed capacity (TIEPI), customer-minutes lost (CML), customer average interruption duration index (CAIDI), customer total average interruption duration index (CTAIDI), customer average interruption frequency index (CAIFI), average service availability index (ASAI), customers experiencing multiple interruptions (CEMIn), energy not supplied (ENS), average energy not supplied (AENS), average customer curtailment index (ACCI), average

M. Čepin, *Assessment of Power System Reliability*,
DOI: 10.1007/978-0-85729-688-7_14, © Springer-Verlag London Limited 2011

system interruption frequency index (ASIFI), average system interruption duration index (ASIDI), average interruption time (AIT), average interruption frequency (AIF), average interruption duration (AID), momentary average interruption frequency index (MAIFI), momentary average interruption event frequency index (MAIFI_E), and customers experiencing multiple sustained interruption and momentary interruption events (CEMSMIn).

14.2.1 System Average Interruption Frequency Index

The SAIFI indicates how often the average customer experiences a sustained interruption over a predefined period of time, usually a year.

$$\text{SAIFI} = \frac{\sum_i N_i}{N_T} \tag{14.1}$$

where the sum is taken over all events i, either at all voltages levels or only at selected ones. N_i is the number of customers interrupted by each incident i, and N_T is total number of customers in the system for which the index is calculated.

SAIFI can also be measured by the mean time between failure (MTBF), which is the reciprocal value of the failure rate λ. SAIFI typical value is mostly between one and two sustained interruptions per year [11]. The value depends on the system configuration and is higher for radial configuration, smaller for underground residential, and the smallest for the grid network [11].

14.2.2 Transformer SAIFI

Transformer SAIFI is used in Finland for SAIFI weighted by the annual energy consumption [2].

14.2.3 Equivalent Number of Interruptions Related to the Installed Capacity

NIEPI is used in Spain as an alternative for SAIFI to quantify the average number of supply interruptions.

$$\text{NIEPI} = \frac{\sum_i P_{\text{ri}}}{P_{\text{rT}}} \tag{14.2}$$

where P_{ri} is the sum of the rating of all interrupted medium-voltage/low-voltage transformers plus the contracted power of all interrupted medium-voltage and high-voltage customers. P_{rT} is the total rating of all medium-voltage/low-voltage transformers plus the total contracted power of all medium-voltage and high-voltage customers connected to the system.

14.2.4 Customer Interruption

Customer interruption is used in United Kingdom instead of SAIFI, but it is expressed as the number of interruptions per 100 customers per year [2].

14.2.5 System Average Interruption Duration Index

SAIDI indicates the total duration of interruption for the average customer during a predefined period of time. It is usually measured in customer-minutes or customer-hours of interruption.

$$\mathrm{SAIDI} = \frac{\sum_i N_i \cdot r_i}{N_T} \tag{14.3}$$

where r_i is the restoration time for each interruption i. Typical values of SAIDI are between 1.5 and 3 h per year [11].

14.2.6 Transformer SAIDI

Transformer SAIDI is used in Finland for SAIDI weighted by the annual energy consumption [2].

14.2.7 Equivalent Interruption Time Related to the Installed Capacity

TIEPI is used in Spain and Portugal to quantify the average time during which the supply to a customer is interrupted.

$$\mathrm{TIEPI} = \frac{\sum_i P_{\mathrm{ri}} \times r_i}{P_{\mathrm{rT}}} \tag{14.4}$$

where P_{ri} is the sum of the rating of all interrupted medium-voltage/low-voltage transformers plus the contracted power of all interrupted medium-voltage and high-voltage customers. P_{rT} is the total rating of all medium-voltage/low-voltage transformers plus the total contracted power of all medium-voltage and high-voltage customers connected to the system.

14.2.8 Customer-Minutes Lost

CML is used in United Kingdom instead of SAIDI [2].

14.2.9 Customer Average Interruption Duration Index

CAIDI represents the average time required to restore service. It is expressed in units of time per interruption, usually in minutes per interruption. From customer point of view, it is closely related with the term mean time to restore or mean time to repair (MTTR) [11].

$$\mathrm{CAIDI} = \frac{\sum_i N_i \times r_i}{\sum_i N_i} = \frac{\mathrm{SAIDI}}{\mathrm{SAIFI}} \tag{14.5}$$

Typical values of CAIDI are between 90 and 150 min per interruption [11]. The value depends on the system configuration and is lower for radial configuration, higher for underground residential, and the highest for the grid network [11].

14.2.10 Customer Total Average Interruption Duration Index

Customer total average interruption duration index represents the total average time in the reporting period that customers who actually experienced an interruption were without power. This index is a hybrid of CAIDI and is similarly calculated except that those customers with multiple interruptions are counted only once. Like SAIDI, it is usually expressed in minutes per customer per year.

$$\mathrm{CTAIDI} = \frac{\sum_i N_i \times r_i}{N_c} \tag{14.6}$$

where N_c is the total number of customers that have experienced at least one interruption during the reporting period.

14.2.11 Customer Average Interruption Frequency Index

CAIFI gives the average frequency of sustained interruptions for those customers experiencing sustained interruptions. The customer is counted once regardless of the number of times interrupted for this calculation. Like SAIFI, it is usually expressed in interruptions per customer per year.

$$\text{CAIFI} = \frac{\sum_i N_i}{N_c} \tag{14.7}$$

14.2.12 Average Service Availability Index

The ASAI represents the fraction of time that a customer has received power during the defined reporting period.

$$\text{ASAI} = 1 - \frac{\sum N_i \times r_i}{N_T \times T} \tag{14.8}$$

where T is the time interval (8,760 or 8,784 h in a leap year)

Another way of looking at ASAI on annual basis is defined with SAIDI, where SAIDI is expressed in hours.

$$\text{ASAI} = \frac{(T - \text{SAIDI})}{T} \tag{14.9}$$

14.2.13 Customers Experiencing Multiple Interruptions

CEMIn indicate the ratio of individual customers experiencing more than n sustained interruptions to the total number of customers served.

$$\text{CEMI}_n = \frac{N_{c,k>n}}{N_T} \tag{14.10}$$

where $N_{c,k\ >\ n}$ is the total number of customers who experienced more than n sustained interruptions and momentary interruption events during the observed period.

14.2.14 Energy Not Supplied

ENS gives the total amount of energy that would have been supplied to the interrupted customers if there would not have been any interruption. It is usually expressed in MWh.

$$\text{ENS} = \sum_i P_i \times r_i = \sum_i E_i \tag{14.11}$$

where P_i is the average load interrupted by each interruption i and E_i is the energy not supplied because of each interruption i.

14.2.15 Average Energy Not Supplied

The AENS index indicates how much energy on average was not served to the customers during a predefined period of time. It is usually expressed in MWh.

$$\text{AENS} = \frac{\sum_i P_i \times r_i}{\sum_i N_i} = \frac{\text{ENS}}{\sum_i N_i} \tag{14.12}$$

14.2.16 Average Customer Curtailment Index

The ACCI indicates how much energy on average was not served to the interrupted customers during a predefine period of time. It is usually expressed in MWh.

$$\text{ACCI} = \frac{\sum P_i \times r_i}{N_i} = \frac{\text{ENS}}{N_i} \tag{14.13}$$

14.2.17 Average System Interruption Frequency Index

The calculation of ASIFI is based on load rather than customers affected. ASIFI is sometimes used to measure distribution performance in areas that serve relatively few customers having relatively large concentrations of load, predominantly industrial/commercial customers. Theoretically, in a system with homogeneous load distribution, ASIFI is the same as SAIFI. It is usually expressed in number of interruptions per year.

$$\text{ASIFI} = \frac{\sum_i P_i}{P_T} \tag{14.14}$$

where P_T is the total rated or contracted power in the system.

14.2.18 Average System Interruption Duration Index

The calculation of ASIDI is based on load rather than customers affected. Its use, limitations, and philosophy are stated in the ASIFI definition. It is usually expressed in minutes per year. Factor 60 can be used for changing unit from hours to minutes.

$$\text{ASIDI} = \frac{60 \times \sum_i P_i \times r_i}{P_T} = \frac{60 \times \text{ENS}}{P_T} \quad (14.15)$$

14.2.19 Average Interruption Time

AIT is a measure for the amount of time that the supply is interrupted. It is similar to ASIDI, which is used in distribution, while AIT is used in transmission. Average interruption time is usually expressed in minutes per year. Factor 60 can be used for changing unit from hours to minutes.

$$\text{AIT} = \frac{60 \times \sum_i E_i}{P_T} = \frac{60 \times \text{ENS}}{P_T} \quad (14.16)$$

where P_T is the average power supplied by the total system, and E_i is the energy not supplied because of each interruption i.

14.2.20 Average Interruption Frequency

AIF is a measure for the number of times per year that the supply is interrupted. It is usually expressed in interruption per customer per year.

$$\text{AIF} = \frac{\sum_i P_i}{P_T} \quad (14.17)$$

where P_i is the power interrupted by each incident i.

14.2.21 Average Interruption Duration

AID is a measure for the average duration of an interruption. It is usually expressed in minutes per interruption. Factor 60 can be used for changing unit from hours to minutes.

$$\text{AID} = \frac{60 \times \sum_i E_i}{\sum_i P_i} = \frac{60 \times \text{ENS}}{\sum_i P_i} \quad (14.18)$$

14.2.22 Momentary Average Interruption Frequency Index

MAIFI indicates the average frequency of momentary interruptions. The upper limit of the duration of a short interruption varies between different countries from 1 to 3 min. MAIFI is usually expressed in number of interruptions per year similarly to SAIFI.

$$\mathrm{MAIFI} = \frac{\sum_i N_i \times N_{IDi}}{N_T} \tag{14.19}$$

where N_{IDi} is the number of interrupting device operations.

14.2.23 Momentary Average Interruption Event Frequency Index

$\mathrm{MAIFI_E}$ indicates the average frequency of momentary interruption events. This index does not include the events immediately preceding a lockout.

$$\mathrm{MAIFI}_E = \frac{\sum_i N_i \times N_{\mathrm{IDE}}}{N_T} \tag{14.20}$$

where N_{IDE} is the interrupting device events during reporting period.

MAIFI typical value is mostly between zero and ten momentary interruptions per year [11]. The value depends on the system configuration and is higher for radial configuration, smaller for underground residential, and the smallest for the grid network, where it is around zero [11].

14.2.24 Customers Experiencing Multiple Sustained Interruption and Momentary Interruption Events

This index is the ratio of individual customers experiencing more than n of both sustained interruptions and momentary interruption events to the total customers served. Its purpose is to help identify customer issues that cannot be observed by using averages.

$$\mathrm{CEMSMIn} = \frac{N_{\mathrm{CT},k>n}}{N_T} \tag{14.21}$$

where $N_{CT,k>n}$ is the total number of customers who have experienced more than n interruptions (sustained and momentary interruptions) during the reporting period.

14.3 Facts About Reliability Indices

The distribution reliability indices are classified into five groups:

- *Sustained interruption indices*: transformer SAIFI, equivalent number of interruptions related to the installed capacity (NIEPI), customer interruption, system average interruption duration index (SAIDI), transformer SAIDI, equivalent interruption time related to the installed capacity (TIEPI), customer-minutes lost (CML), customer average interruption duration index (CAIDI), customer total average interruption duration index (CTAIDI), customer average interruption frequency index (CAIFI), and average service availability index (ASAI), customers experiencing multiple interruptions (CEMIn)
- *Energy-based indices*: energy not supplied (ENS), average energy not supplied (AENS), and average customer curtailment index (ACCI)
- *Load-based indices*: average system interruption frequency index (ASIFI) and average system interruption duration index (ASIDI)
- *Indices for transmission system*: average interruption time (AIT), average interruption frequency (AIF), and average interruption duration (AID)
- *Indices for short interruption*: momentary average interruption frequency index (MAIFI), momentary average interruption event frequency index ($MAIFI_E$), and customers experiencing multiple sustained interruption and momentary interruption events (CEMSMIn)

The reliability indices, SAIFI, SAIDI, and CAIDI, are the main indices used in the majority of countries. These indices are defined among others in standard [1], where weighting based on number of customers is used. With both indices, SAIFI and SAIDI, a reduction in value indicates an improvement in the continuity of supply. With CAIDI this is not the case: A reduction of both SAIDI and SAIFI could still result in an increase of CAIDI. Whereas CAIDI remains a useful index, it is not suitable for comparisons or for trend analysis.

To adequately measure performance, both duration and frequency of customer interruptions must be examined at various system levels. The most commonly used indices for measuring performance are SAIFI, SAIDI, CAIDI, and ASAI. All of these indices provide information about average system performance. Many utilities also calculate indices on a feeder basis to provide more detailed information for decision making. Averages give general performance trends for the utility, however, the use of averages leads to loss of detail that could be critical to decision making. For example, using system averages alone will not provide information about the interruption duration experienced by any specific customer. It is difficult for most utilities to provide information on a customer basis. The tracking of specific details including specific interruptions and averages is accomplished by improving tracking capabilities.

Table 14.1 Support for example calculation of reliability indices

Outage identification	No. of customers	Duration (min)	Customer-hours
1	10	30	5
2	100	10	16.67
3	1	75	1.25
4	2	60	2
All	113		24.92

14.4 Example

The example relates to calculation of reliability indices for the specified time period, where four outages have been recorded for the utility. The utility has a total of 10,000 customers. Table 14.1 shows information for each outage including date and time of occurrence of the event, number of customers affected, duration of the event, and calculated customer-hours obtained as a product of the last two.

There were 113 customers interrupted during four separate events and total number of customers served by utility is 10,000.

$$\mathrm{SAIFI} = \frac{113}{10,000} = 0.0113 \tag{14.22}$$

It can be seen from Table 14.1 that the first outage was affecting 10 customers, which were out of service for 30 min, which equals to 0.5 h. Therefore, the customer-hours are 10 × 0.5 or 5 customer-hours. The customer-hours are calculated for each outage and then they are summed for a total of 24.92 customer-hours or 1,495 customer-minutes.

The calculation of SAIDI is simple and it gives 0.1495 min.

$$\mathrm{SAIDI} = \frac{1,495}{10,000} = 0.1495 \text{ min.} \tag{14.23}$$

This says that the average customer was out for approximately 0.15 min. If SAIDI is calculated for each day, the monthly SAIDI is calculated by sum of the daily values.

The customer-minutes are 1,495 and 113 customers were interrupted. Therefore, the calculation of CAIDI gives 13.23 min. On average, any customer who experienced an outage was out of service for 13.23 min.

$$\mathrm{CAIDI} = \frac{1,495}{113} = 13.23 \text{ min.} \tag{14.24}$$

The customers at this utility had a probability of 0.0113 of experiencing a power outage. SAIFI can also be found by dividing the SAIDI value by the CAIDI value.

$$\mathrm{SAIFI} = \frac{\mathrm{SAIDI}}{\mathrm{CAIDI}} = \frac{0.1495}{13.23} = 0.0113 \tag{14.25}$$

CAIFI is evaluated as ratio between four events and 113 customers interrupted and gives the average number of interruptions for a customer who was interrupted.

$$\mathrm{CAIFI} = \frac{4}{113} = 0.035 \quad (14.26)$$

References

1. IEEE Std 1366 (2003) IEEE guide for electric power distribution reliability indices. IEEE
2. Council of European Energy Regulators (2008) The 4th benchmarking report on quality of electricity supply
3. Kueck JD, Kirbly BJ, Overhalt PN et al (2004) Measurement practices for reliability and power quality: a toolkit of reliability measurement practices. Oak Ridge National Laboratory
4. Layton L (2004) Electric system reliability indices. http://wwwl2engcom/. Accessed 25 Nov 2010
5. Murray J (2008) Six-year electric service reliability statistics summary 2002–2007. Oregon Investor-Owned Utilities. http://www.puc@state.or.us. Accessed 26 Nov 2010
6. Billinton R, Allan R (1996) Reliability evaluation of power systems. Plenum, New York
7. Dhillon BS (2007) Applied reliability and quality fundamentals methods and procedures. Springer, London
8. Rietz R, Sen PK (2006) Costs of adequacy and reliability of electric power. IEEE 525–529
9. Volkanovski A (2008) Impact of offsite power system reliability on nuclear power plant safety. PhD thesis, University of Ljubljana
10. Burke JJ (1994) Power distribution engineering: fundamentals and applications. Marcel Dekker, New York
11. Short TA (2006) Distribution reliability and power quality. Taylor & Francis, Boca Raton
12. Settembrini RC, Fisher JR, Hudak NE (1991) Reliability and quality comparisons of electric power distribution systems. IEEE Power Engineering Society, transmission and distribution conference
13. Christie RD (2002) Statistical methods of classifying major event days in distribution systems, IEEE Power Engineering Society summer meeting
14. Brown RE (2006) Power system reliability, Chapter 19. In: Grigsby LL (ed) Power systems. Taylor & Francis
15. Philipson L, Willis HL (2006) Understanding electric utilities and de-regulation. Taylor & Francis, Boca Raton, FL
16. Willis HL (2003) Spatial electric load forecasting. Marcel Dekker, New York
17. Casazza J, Delea F (2003) Understanding electric power systems: an overview of the technology and the marketplace. Wiley, Hoboken, NJ
18. Sheblé GB (2001) Power system planning (reliability). In: Grigsby LL (ed) The electric power engineering handbook. CRC, Boca Raton, FL
19. Ward DJ (2006) Power distribution: standard handbook for electrical engineers. McGraw-Hill, New York
20. Vidya Sagar E, Prasad PVN (2010) Reliability evaluation of automated radial distribution system. IEEE PMAPS, pp 558–563
21. Viet NH, Ban NV (2010) Automation and reliability improvement for distribution feeder. IEEE PMAPS, pp 609–614

22. Li W, Wangdee W, Choudhury P (2010) A probabilistic planning approach to single circuit supply systems. IEEE PMAPS, pp 691–696
23. Grigsby LL (2006) Power systems. Taylor & Francis
24. Chowdhury A, Koval D (2009) Power distribution system reliability: practical methods and applications. Wiley, Hoboken, NJ

Chapter 15
Power System Reliability Method

To kill an error is as good a service as, and sometimes even better than, the establishing of a new truth or fact
Charles Darwin

15.1 Introduction

The electric power system is among the most complex systems ever known. It consists of uncounted number of facilities and structures, systems and subsystems, components and equipment, and the complex interactions among all those [1–13]. Therefore, it is difficult to assess its reliability as one system measure. If the system is considered as a static system and its components as entities of the system, which are not represented by their dynamic behavior, the system reliability can be assessed. The electric power system reliability can be assessed based on the configuration of the system, based on the reliability of the components of the system and based on the viewpoint of power delivery to the loads of the power system [4–7]. Example evaluations of small systems support the development of the method [8–11]. Example evaluations of larger systems support the applicability and improvement of the method [6, 7].

Many issues are connected to the power system reliability because of its complexity. One of the issues is the safety of nuclear power plants, which largely contribute to high reliability of power system. And high reliability of power system contributes to high level of nuclear power plant safety [6, 14–22]. The other issue is the vulnerability of strategic infrastructure because of similar topology and some similar operational characteristics [23–27]. Because of many related issues, the information background of the field of power system reliability is large, including existing methods and analyses of power systems [28–31], maintenance and outages and optimizations related to power systems and substations [32–37], risk and reliability methods applied in related fields [38–45], reliability measures [46, 47], behavior of power systems [48–51], data supporting the reliability analyses [51–57], and consideration of common cause failures [57].

M. Čepin, *Assessment of Power System Reliability*,
DOI: 10.1007/978-0-85729-688-7_15, © Springer-Verlag London Limited 2011

Fig. 15.1 Example power system 3NET with three buses, three generators and three loads

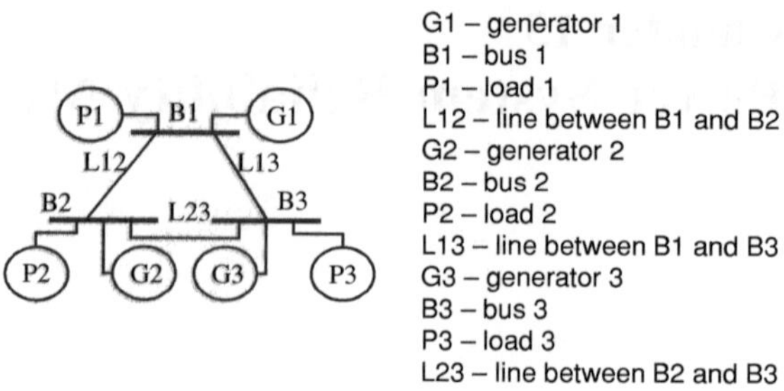

15.2 Definition of the Power System Reliability

The reliability of the power system R_{PS} is defined from its complement, i.e., unreliability U_{PS} [4–7].

$$R_{PS} = 1 - U_{PS} = 1 - \sum_{i=1}^{NL} U_i \frac{K_i}{K} \tag{15.1}$$

$$U_{PS} = \sum_{i=1}^{NL} U_i \frac{K_i}{K} \tag{15.2}$$

where R_{PS} is the power system reliability, U_{PS} is the system unreliability, U_i is the unreliability of the power delivery to the ith load, NL is the number of loads in system, K_i is the size of ith load (MW), K_i/K is the weighting factor for ith load, and K is the complete load of the power system represented as the sum of all the loads.

Unreliability of the power delivery to the ith load can be assessed as the top-event probability of the respective fault tree analysis. Consideration of each of the loads can be done as consideration of a subsystem. Evaluation of subsystems represents the input to the evaluation of the overall power system. The unreliabilities of power delivery to the loads of the system are considered as weighted to get the overall measure of the power system reliability [4, 5].

$$K = \sum_{i=1}^{NL} K_i \tag{15.3}$$

The weights are normalized according to the power of the power system as a whole.

For example, Fig. 15.1 shows an example power system 3NET with three buses, three generators, and three loads. The losses are neglected. The sum of power of three generators equals to the sum of the loads (Tables 15.1 and 15.2).

The following section gives the method description including how the unreliabilities of the subsystems or the unreliabilities of the system from the viewpoint of one specific load are calculated.

Table 15.1 Quantitative results of subsystems of power system 3NET

Subsystem identification	Unreliability	Unreliability×weighting factor	Weight	Load size
Failure of power delivery to load P1 on bus B1	2.11E-6	5.28E-7	2.5E-1	300 MW
Failure of power delivery to load P2 on bus B2	4.21E-6	1.40E-6	3.33E-1	400 MW
Failure of power delivery to load P3 on bus B3	6.32E-6	2.63E-6	4.17E-1	500 MW

Table 15.2 Quantitative results of overall power system 3NET

System Identification	Unreliability	Unreliability (weighted)	Load size
3NET	4.21E-6	4.56E-6	1200 MW

15.3 Method Description

The method for power system reliability evaluation includes the evaluation of the system from the viewpoint of each of the loads of the system [5, 7–10]. The model of unreliability of the specific load of the system is in theory unique for each load. Therefore, the evaluation of the system consists of evaluations of the subsystems and of the overall consideration of the obtained results to the system reliability evaluation. The system is modeled and the system models from viewpoints of specific loads differ between themselves and represent the subsystem models. The results of those subsystem models are then used in the power system reliability calculation.

The method is developed in a way that the increase of the power system under consideration does not require significant or even unsolvable increase of the model. Increase of the power system means larger number of nodes or/and larger number of connections between the nodes or/and more complex models of nodes.

The power system is assumed as a system consisting of a number of nodes (e.g., buses, switchyards, and transformer stations), which are connected with connections (e.g., power lines and power cables). Each node can be connected to loads. Each node can be connected to an electric power supply (e.g., generator or generators; although it can be connected to other power system or to other part of the network).

The prerequisite for the method development is the representation of the system topology. The nodes of the network and their connections are represented by the buses in the power system and the power lines between the buses. When the system topology is defined, the model of power flow paths is developed. When the model of power flow paths is developed, fault tree is constructed and analyzed.

Interpretation of the fault tree analysis results includes the power system reliability calculated from its complement, i.e., unreliability and the importance measures, which identify the most important components of the system from various viewpoints.

15.3.1 Model of the System Topology

The representation of the system topology is performed through the adjacency matrix or through the matrix of connections.

15.3.1.1 Adjacency Matrix

The adjacency matrix is a matrix with the rows and columns labeled with a 1 or 0 in position (v_{ij}) according to whether the bus i is connected to bus j directly or not. The rows follow the running index i of buses. The columns follow the running index j of buses. If the value of position v_{ij} is 0, the buses i and j are not directly connected. If the value of position v_{ij} is 1, the buses i and j are directly connected. The diagonal elements of the matrix equal 0, as the bus is not connected to itself. Equation shows an example of adjacency matrix, for example, system 3NET.

$$A = \begin{bmatrix} 0 & 1 & 1 \\ 1 & 0 & 1 \\ 1 & 1 & 0 \end{bmatrix} \tag{15.4}$$

The matrix is full of values 1 because every bus is connected to both other two buses. The first row of the adjacency matrix shows that the bus 1 is connected to bus 2 and to bus 3 as the number 1 appears at both positions in the second and the third column of the row 1. The second row of the adjacency matrix shows that the bus 2 is connected to bus 1 and to bus 3 as the number 1 appears at both positions in the first and the third column of the row 2. The third row of the adjacency matrix shows that the bus 3 is connected to bus 1 and to bus 2 as the number 1 appears at both positions in the first and the second column of the row 3.

The matrix is symmetric and the information of connection between the buses is written in the matrix two times. The triangular part of the matrix would be sufficient, because if the bus i is directly connected to bus j, also the vice versa is true. Or, if the bus i is not directly connected to bus j, also the vice versa is true.

15.3.1.2 Matrix of Connections

The matrix of connections is a matrix with the rows and columns labeled with identifying numbers of buses. It identifies the buses, which are connected to other buses and buses, which are not connected to others.

The first number in a row is a sequential number of a specific bus. The next numbers identify the sequential numbers of buses, which are connected to the bus identified in the first column of the row. If some buses are not connected to the bus identified in the first column of the row, values 0 appear at the respective locations. The matrix is full of values because every bus is connected to all other three buses.

$$A = \begin{bmatrix} 1 & 2 & 3 \\ 2 & 1 & 3 \\ 3 & 1 & 2 \end{bmatrix} \tag{15.5}$$

The first row of the matrix shows that the bus 1 is connected to buses 2 and 3. The second row shows that the bus 2 is connected to buses 1 and 3. The third row shows that the bus 3 is connected to buses 1 and 2.

15.3.2 Model of Power Flow Paths

The model of power flow paths can be realized with the functional tree of power flow paths or with the rooted tree. The functional tree of power flow paths is simpler. The rooted tree gives additional information to the model of the system about power flows through power lines and measured voltage level at flow paths. This information about power flows through power lines and measured voltage level at flow paths is used in a later step of fault tree development.

15.3.2.1 Functional Tree of Power Flow Paths

The functional tree of power flow paths is a figure of flow paths from the power source to the loads. All possible power flow paths between the power source and the loads are identified, using developed recursive procedure for formation of functional tree of power flow paths, which is related to the topology of the system. The recursive procedure is standard recursion with the marking of passed nodes to avoid closed cycles.

Figure 15.2 shows functional trees of power flow paths 3NET for all three loads. The abbreviations B1, B2, and B3 represent the busses from the Figure 15.1: bus-1, bus-2, and bus-3, respectively. The functional tree of power flow paths for the load P1 starts with the bus B1, because the load P1 is connected to the powers source in the bus B1. Three sources of power delivery to bus B1 are identified in the second step: generator G1 connected to the bus B1, the line between bus B1 and bus B2, which is marked with L12 and the line between bus B1 and bus B3, which is marked with L13. The generator G1 is the endpoint of the branch of the functional tree of power flow paths, because generators are the sources of power delivery. For the line L12 and for the line L13, the functional tree of power flow paths is further developed. The source of line L12 in the

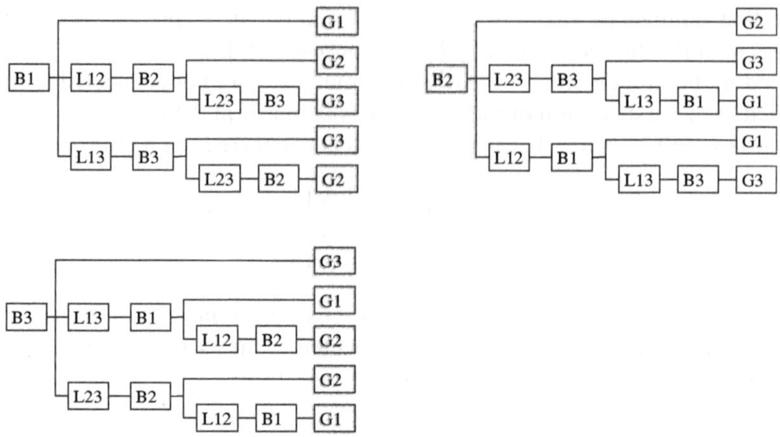

Fig. 15.2 Functional trees of power flow paths 3NET for all three loads

direction of bus B1 is the bus B2, which is the next component in the branch of the functional tree of power flow paths. Similarly, the source of line L13 in the direction of bus B1 is the bus B3, which is the next component in the branch of the functional tree. Both continuations of the development of the functional tree of power flow paths are described in the following two paragraphs.

Two sources of power delivery are identified in the next step for ensuring the power delivery to bus B2: generator G2 connected to the bus B2 and the line between bus B2 and bus B3, which is marked with L23. The generator G2 is the endpoint of the branch of the functional tree. For the line L23, the functional tree of power flow paths is further developed. The source of line L23 in the direction of bus B2 is the bus B3, which is the next component in the branch of the functional tree of power flow paths. Only one source of power delivery is identified in the next step for ensuring the power delivery to bus B3: generator G3 connected to the bus B3. This source of power delivery is the only one, because the connection with the bus B1, which exists, and bus B1, which is behind this connection, is not considered at this point because of avoiding the looping of the model. Namely, the bus B1 and its possible failure consequently have been considered at the start of the development of the functional tree.

Two sources of power delivery are identified for ensuring the power delivery to bus B3: generator G3 connected to the bus B3 and the line between bus B2 and bus B3, which is marked with L23. The generator G3 is the endpoint of the branch of the functional tree. For the line L23, the functional tree of power flow paths is further developed. The source of line L23 in the direction of bus B3 is the bus B2, which is the next component in the branch of the functional tree of power flow paths. Only one source of power delivery is identified in the next step for ensuring the power delivery to bus B2: generator G2 connected to the bus B2. This source of power delivery is the only one, because the connection with the bus B1, which exists, and bus B1, which is behind this connection, is not considered at this point

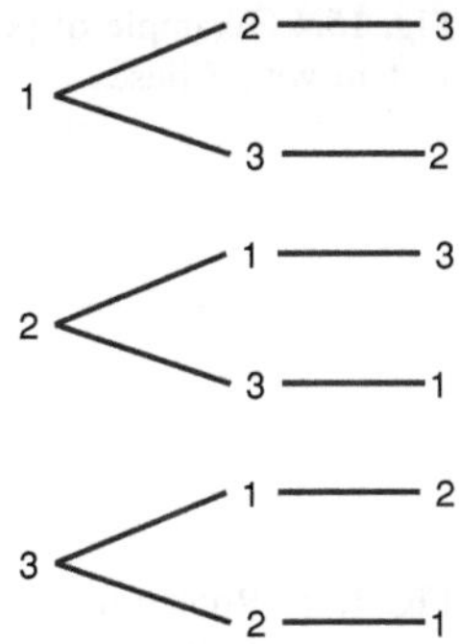

Fig. 15.3 Rooted trees of power flow paths 3NET for all three loads

because of avoiding the looping of the model. Namely, the bus B1 and its possible failure consequently have been considered at the start of the development of the functional tree.

The development of the functional tree of power flow paths for the load P2, which starts with the bus B2, is similar to the presented development of the functional tree for the load P1, which starts with the bus B1. The development of the functional tree of power flow paths for the load P3, which starts with the bus B3, is similar.

15.3.2.2 Rooted Tree

The rooted tree is a figure of flow paths from the power source to the loads. A recursive procedure for composition of rooted tree of power flow paths between the power source and the loads is developed. The recursive procedure outcome is the rooted tree, which is related to the topology of the system. The recursive procedure uses standard recursion with marking of the considered nodes to avoid closed cycles. Figure 15.3 shows rooted trees of power flow paths 3NET for all three loads.

The identified power flow paths of energy delivery between nodes of the network or in other words between substations of the power system are tested for consistency.

- *Test of overloaded line* If there is an overloaded line in the flow path, then that flow path is marked as overloaded.
- *Test of required voltage* If there is substation with the decreased voltage, which exceeds the allowed voltage drop, the related flow paths are marked as decreased voltage.

The prerequisite of those two tests is the integration of the rooted tree development with the method for the power flow analysis of the specific system under consideration [6, 7]. The direct current power flow model was used in references

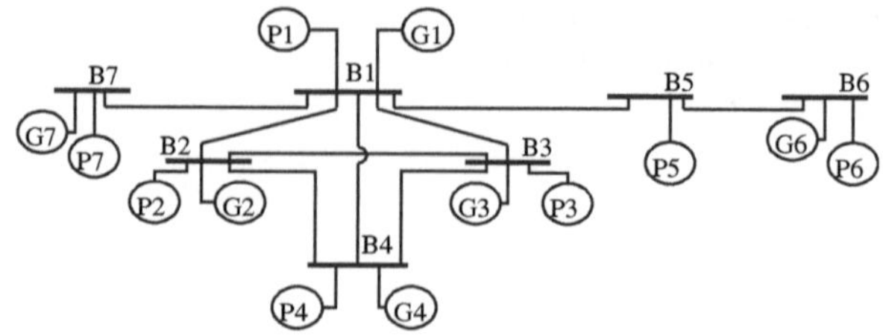

Fig. 15.4 Example of power system with 7 buses

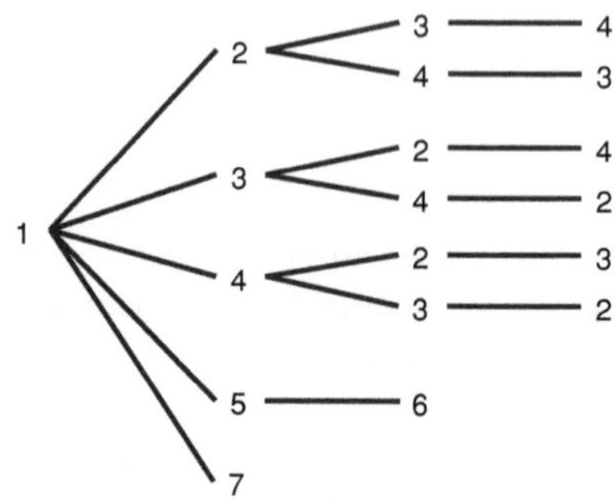

Fig. 15.5 Rooted tree of power flow paths of power system with 7 buses for load 1

[6, 7] for such purposes. The direct current power flow model was obtained from the alternating current model of power system approximating that voltages in all buses are equal to nominal, considering the differences of angles of voltages are very small and neglecting the losses in power system [6, 7]. The direct current power flow model gives a linear relationship between the power flowing through the lines and the power input at the nodes [6, 7]. The direct current power flow method is simpler, but it gives less accurate results. The other methods that can be used for the power flow analysis are presented in Chap. 11.

The related features of the rooted tree technique are further described on a larger example. Figure 15.4 shows an example of power system with seven buses.

Figure 15.5 shows the rooted tree of power flow paths of power system with seven buses for load 1. Tests of overloaded lines and tests of required voltages were either not performed or they reveal no change to the rooted tree as no overloaded lines and no excess voltage drops are identified.

Figure 15.6 shows three cases of the rooted tree for the same system. The *left variant* shows the case where the voltage decrease at bus 5 is more than allowed limit (e.g., decrease of voltage is more than 5%). In such case, the line between buses 1 and 5 and the line between 5 and 6 are represented with *dashed lines*. The *middle variant* (Fig. 15.6) shows the case where the power capacity of the generator 7 is smaller than the required power of load 1. The line between bus 1 and 7 is represented with a *dashed line*, because the generator 7 alone is not capable of supplying the load 1. The *right variant* shows both mentioned cases as the third case.

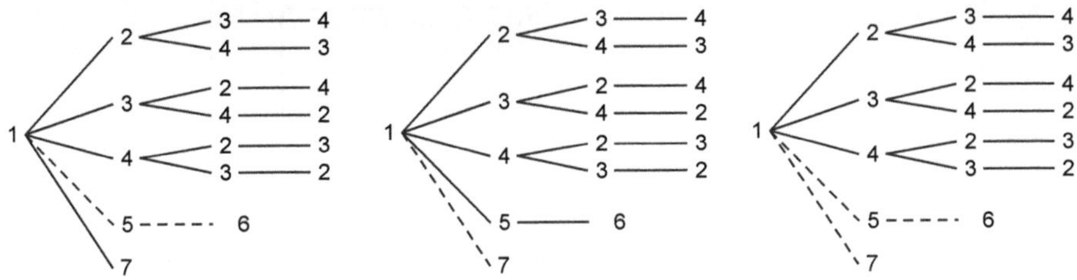

Fig. 15.6 Rooted tree variants

When the rooted trees are developed, the fault tree development and analysis take place. There are options for considerations of rooted trees in the fault tree evaluation.

The simplest option is the one with consideration of the basic rooted tree without tests of overloaded lines and voltage drops. This means that the *dashed lines* in the rooted trees are neglected.

The second option is consideration of the overloaded lines and voltage drops with significantly increased failure probabilities of related equipment.

The third option is the most conservative, which means that the overloaded lines and buses with voltage drops are even not considered as possible sources of power delivery in the system topology at all. Further information on the consistency tests and details on the rooted tree construction is presented in references [6, 7].

15.3.3 Fault Tree Development

The model for the unreliability evaluation of the system from the viewpoint of specific load can be performed with the fault tree analysis. Consideration of common cause failures is optional feature within the method. The fault tree analysis is performed for all the loads of the system, which are three for the 3NET example system. The first fault tree connected with the failure of power delivery to load P1 starts with definition of the respective top event, which is the failure of power delivery to load P1. The second fault tree starts with top event, which is the failure of power delivery to load P2. The third fault tree starts with top event, which is the failure of power delivery to load P3. Because the system is completely symmetric, all the fault trees are similar to each other. They differ only in gates and basic events descriptions, which are related to specific line, bus, or generator, as required by the system and fault tree structure. Because of their similarities, only one fault tree is described in more details. The construction of other two is similar.

The fault tree for the failure of the power delivery to load P1 starts with the top event. The top event represents the failure of power delivery to the load P1. Figure 15.7 shows the starting part of the fault tree, whose development goes hand in hand with the development of the functional tree of power flow paths.

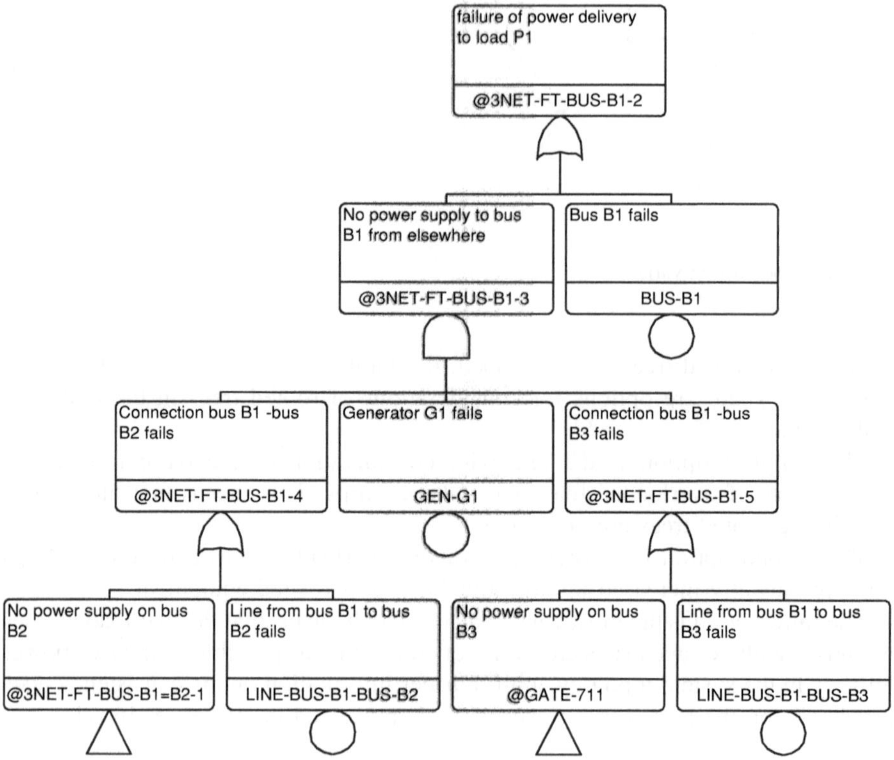

Fig. 15.7 Fault tree for the failure of power delivery to load P1 (from top event)

The functional tree of power flow paths for the load P1 starts with the bus B1, so all failures of bus B1 are the basic event, which can cause the top event. If the bus B1 fails, the power delivery to load P1 is lost. The other event, which can fail the power delivery to load P1, is the failure of power delivery to bus B1.

The functional tree of power flow paths for the load P1 shows that three sources of power delivery to bus B1 are identified. All three have to fail in order that the power delivery fails, which means that AND gate connects all three related events: all failures of generator G1 connected to the bus B1, all failures of the power line L12 and its inputs, and all the failures of the line L13 and its inputs. All failures of generator G1 are represented by basic event, because the generator is the source of power delivery. The other two events are OR gates, because several ways exist in which each of the mentioned failures can occur. The OR gate: failures of the power line L12 and its inputs (shortly written in the fault tree: connection bus B1–bus B2 fails) connect the line L12 failure as the basic event (shortly written in the fault tree: line from bus B1–bus B2 fails) and the OR gate representing the failure of inputs to line L12 (shortly written in the fault tree: no power supply on bus B2). This OR gate is marked as triangle, which means that the fault tree is continued in the next figure. The information, that this gate is OR gate, is presented in the

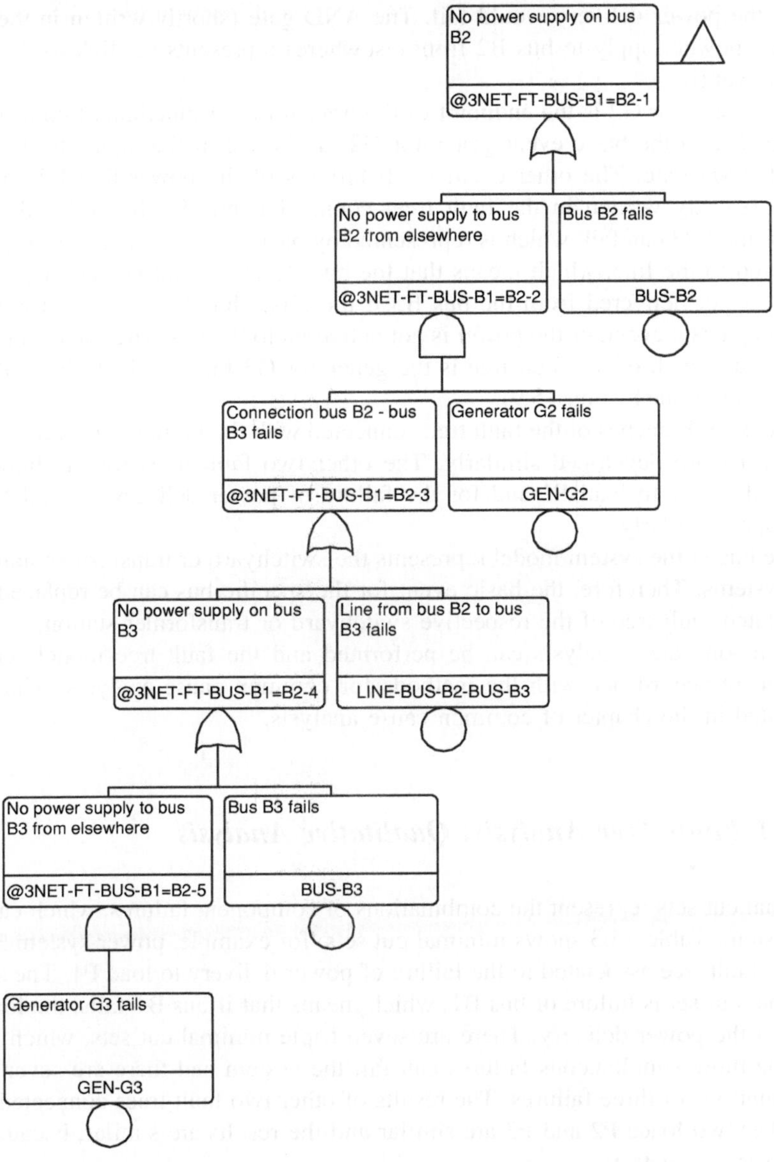

Fig. 15.8 Fault tree for the failure of power delivery to load P1 (continuation 1)

following figure (Fig. 15.8), which starts with the event that has been continued, where the OR gate is the starting gate. This event can occur if the bus B2 fails, which is marked as basic event, or if no power supply is provided (shortly written in the fault tree: no power supply to bus B2 from elsewhere).

The functional tree of power flow paths shows that two sources of power delivery are identified for ensuring the power delivery to bus B2 and both have to fail if the power delivery would fail. The AND gate (shortly written in the fault tree: no power supply to bus B2 from elsewhere) represents the link to the mentioned events.

The generator G2 is the endpoint of the branch of the functional tree of power flow paths, so the basic event generator G2 fail is one of the inputs to the mentioned AND gate. The other event is all failures of the power line L23 and its inputs (shortly written in the fault tree: connection bus B2–bus B3 fails). The power line L23 can fail, which is represented by basic event, or no power supply is delivered to the line, which means that the bus B3 is without power supply. The power is not delivered from the bus B3 if there is a bus failure, which is represented by basic event, or the power is not delivered to the bus. The last event in the presented branch of the fault tree is the generator G3 failure. Here, the fault tree ends at this branch (Fig. 15.9).

The other branches of the fault tree connected with the failure of power delivery to load P1 are developed similarly. The other two fault trees for the failure of power delivery to load P2 and for the failure of power delivery to load P3 are developed similarly.

The bus in the system model represents the switchyard or transformer station in real systems. Therefore, the basic event for the specific bus can be replaced with the related fault tree of the respective switchyard or transformer station.

Common cause analysis can be performed and the fault tree models can be updated in accordance with the methods for common cause analysis, which are presented in the chapter of common cause analysis.

15.3.4 Fault Tree Analysis: Qualitative Analysis

Minimal cut sets represent the combinations of component failures, which can fail the system. Table 15.3 shows minimal cut sets, for example, power system 3NET for the fault tree associated to the failure of power delivery to load P1. The single minimal cut set is failure of bus B1, which means that if bus B1 fails, the load P1 has lost the power delivery. There are seven triple minimal cut sets, which mean that the three simultaneous failures can fail the system and there are seven such combinations of three failures. The results of other two fault trees connected with the other two loads P2 and P3 are similar and the results are similar, because the system is symmetric.

15.3.5 Fault Tree Analysis: Quantitative Analysis

The prerequisite for the quantitative analysis is collection of data about the failure probabilities of modeled equipment. Sources of data bases include IEEE Standard 500 [54] and IAEA-TECDOC-478 [56]. Quantitative fault tree analysis includes

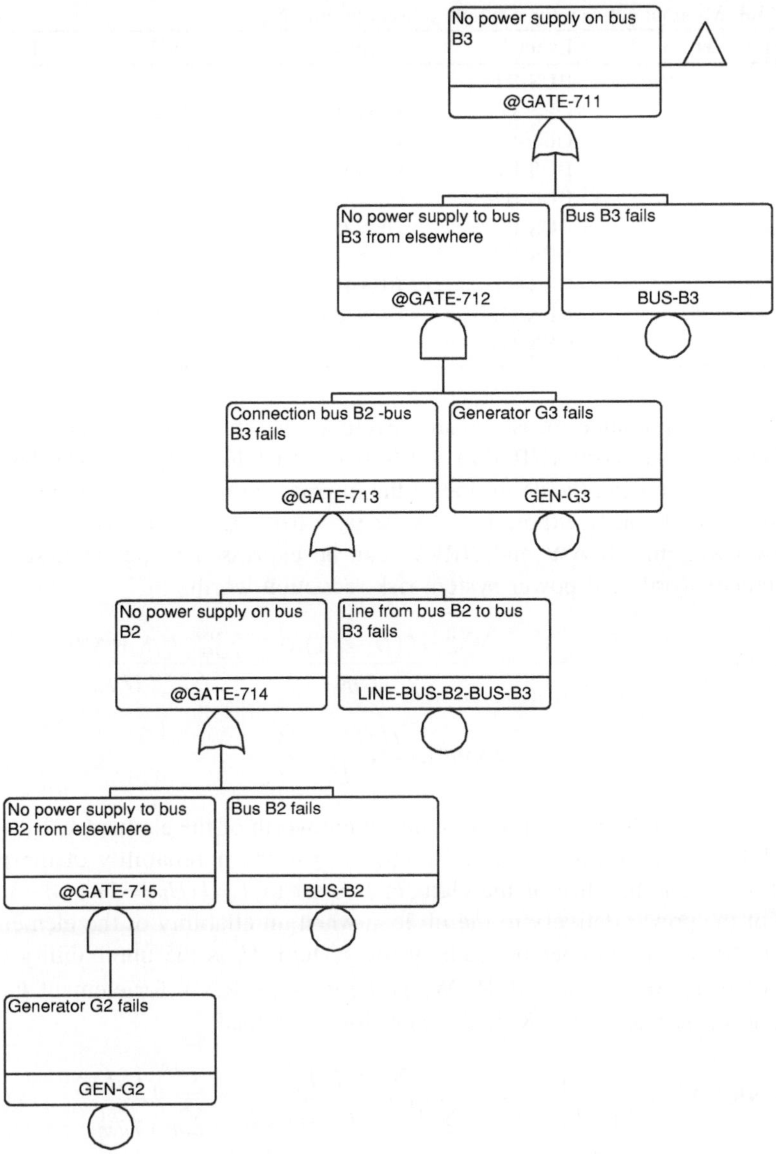

Fig. 15.9 Fault tree for the failure of power delivery to load P1 (continuation 2)

minimal cut sets and failure probabilities, which gives the system unreliability, and importance measures. The importance measures can be calculated based on a single fault tree evaluation as shown in the chapter of the fault tree analysis or they can use novel equations, which are developed given that the network components importance measures can be better evaluated.

Table 15.3 Minimal cut sets for example power system 3NET

Minimal cut set no.	Event 1	Event 2	Event 3	Event 4
1	BUS-B1			
2	GEN-G1	LINE-L12	LINE-L13	
3	GEN-G1	GEN-G2	GEN-G3	
4	BUS-B2	GEN-G1	GEN-G3	
5	BUS-B2	GEN-G1	LINE-L13	
6	BUS-B3	GEN-G1	LINE-L12	
7	BUS-B3	GEN-G1	GEN-G2	
8	BUS-B2	BUS-B3	GEN-G1	
9	GEN-G1	GEN-G3	LINE-L12	LINE-L23
10	GEN-G1	GEN-G2	LINE-L13	LINE-L23

New risk importance measures are developed for the power system: Network Risk Achievement Worth (NRAW) and Network Risk Reduction Worth (NRRW). They are defined using the definition of the importance measures for fault tree and the system unreliability expression. As the term network is a descriptive term for the power system, NRAW and NRRW can be expressed as power system risk achievement worth and power system risk reduction worth.

$$\mathrm{NRAW}^k = \frac{U_{\mathrm{PS}}(U_k = 1)}{U_{\mathrm{PS}}} = \frac{\sum_{i=1}^{\mathrm{NL}} U_i(U_k = 1)K_i}{\sum_{i=1}^{\mathrm{NL}} U_i K_i} = \frac{\sum_{i=1}^{\mathrm{NL}} U_i K_i \mathrm{RAW}_i^k}{\sum_{i=1}^{\mathrm{NL}} U_i K_i} \tag{15.6}$$

$$\mathrm{RAW}_i^k = \frac{U_i(U_k = 1)}{U_i} \tag{15.7}$$

where NRAW^k is the network risk achievement worth of the element k, U_{PS} is the unreliability of the power system, $U_{\mathrm{PS}}(U_k = 1)$ is the unreliability of the power system when unreliability of the element k is set to 1, $U_i(U_k = 1)$ is the unreliability of the power delivery to the ith load when unreliability of the element k is set to 1, NL is the number of loads in the system, U_i is the unreliability of the power delivery to the ith load, RAW_i^k is the value of RAW for element k corresponding to the load i, and K_i is the capacity of ith load.

$$\mathrm{NRAW}^k = \frac{U_{\mathrm{PS}}}{U_{\mathrm{PS}}(U_k = 0)} = \frac{\sum_{i=1}^{\mathrm{NL}} U_i K_i}{\sum_{i=1}^{NL} U_i(U_k = 0)K_i} = \frac{\sum_{i=1}^{\mathrm{NL}} U_i K_i}{\sum_{i=1}^{\mathrm{NL}} \frac{U_i K_i}{PRW_i^k}} \tag{15.8}$$

$$\mathrm{PRW}_i^k = \frac{U_i}{U_i(U_k = 0)} \tag{15.9}$$

where NRRW^k is the network risk reduction worth of the element k, $U_{\mathrm{PS}}(U_k = 0)$ is the unreliability of the power system when unreliability of element k is set to 0, $U_i(U_k = 0)$ is the unreliability of the power delivery to the ith load when unreliability of element k is set to 0, and RAW_i^k is the value of the RRW for element k corresponding to the load i.

Table 15.4 Minimal cut sets and failure probabilities, for example, power system 3NET

Minimal cut set no.	Failure probability	%	Event 1	Event 2	Event 3	Event 4
1	1.00E-09	47.6	GEN-G1	LINE-L12	LINE-L13	
2	1.00E-09	47.6	GEN-G1	GEN-G2	GEN-G3	
3	1.00E-10	4.76	BUS-B1			
4	1.00E-12	0.05	GEN-G1	GEN-G3	LINE-L12	LINE-L23
5	1.00E-12	0.05	GEN-G1	GEN-G2	LINE-L13	LINE-L23
6	1.00E-16	0	BUS-B2	GEN-G1	GEN-G3	
7	1.00E-16	0	BUS-B2	GEN-G1	LINE-L13	
8	1.00E-16	0	BUS-B3	GEN-G1	LINE-L12	
9	1.00E-16	0	BUS-B3	GEN-G1	GEN-G2	
10	1.00E-23	0	BUS-B2	BUS-B3	GEN-G1	

15.3.6 Fault Tree Analysis Results

Tables 15.4 and 15.5 show the quantitative analysis for example power system 3NET for the fault tree associated to the failure of power delivery to load P1. The fictitious set of used data considering failure probabilities of components is given in the third column from the left in Table 15.5. Table 15.4 gives the minimal cut sets and their failure probabilities, for example, power system 3NET.

Table 15.5 gives the failure probabilities and importance factors for the components of the system. The failure probabilities used for the simple example system are fictitious values and do not represent the real probabilities of real components.

The results of the other two fault trees associated to the other two loads P2 and P3 are similar and the results are similar, because the system is symmetric.

15.4 Larger and Real Examples

Larger examples have been performed [6, 7]. Application of the large examples shows that the method is applicable to the real systems. The method was applied to a standardized power system named IEEE RTS-96 [6, 7]. The method was applied to an example of regional power system. Figure 15.10 shows the example system [6, 7].

The system consists of 19 substations and 25 interconnections, which include 10 transformers and 15 power lines, with 13 substations directly connected to loads and 10 substations directly connected to generators. The common cause failures are considered for 12 interconnections. The power system is constructed on the basis of the system configuration presented in reference [13].

The 220- and 400-kV lines are considered in the power system configuration, including two 110-kV lines connected to the power plants in substations Šoštanj and Brestanica. The 110-kV connection to the substation Brestanica is included in the model because it can be connected directly to the nuclear power plant Krško as

Table 15.5 Failure probabilities and importance factors: single fault tree quantification

No	Basic event	Failure probability	Fussel–Vesely importance	Fractional contribution	Risk decrease factor	Risk increase factor
1	GEN-G1	1.00E-03	9.52E-01	9.52E-01	2.10E+01	9.52E+02
2	LINE-L13	1.00E-03	4.76E-01	4.76E-01	1.91E+00	4.77E+02
3	GEN-G3	1.00E-03	4.76E-01	4.76E-01	1.91E+00	4.77E+02
4	GEN-G2	1.00E-03	4.76E-01	4.76E-01	1.91E+00	4.77E+02
5	LINE-L12	1.00E-03	4.76E-01	4.76E-01	1.91E+00	4.77E+02
6	BUS-B1	1.00E-10	4.76E-02	4.76E-02	1.05E+00	4.76E+08
7	LINE-L23	1.00E-03	9.51E-04	9.51E-04	1.00E+00	1.95E+00
8	BUS-B2	1.00E-10	9.51E-08	9.51E-08	1.00E+00	9.52E+02

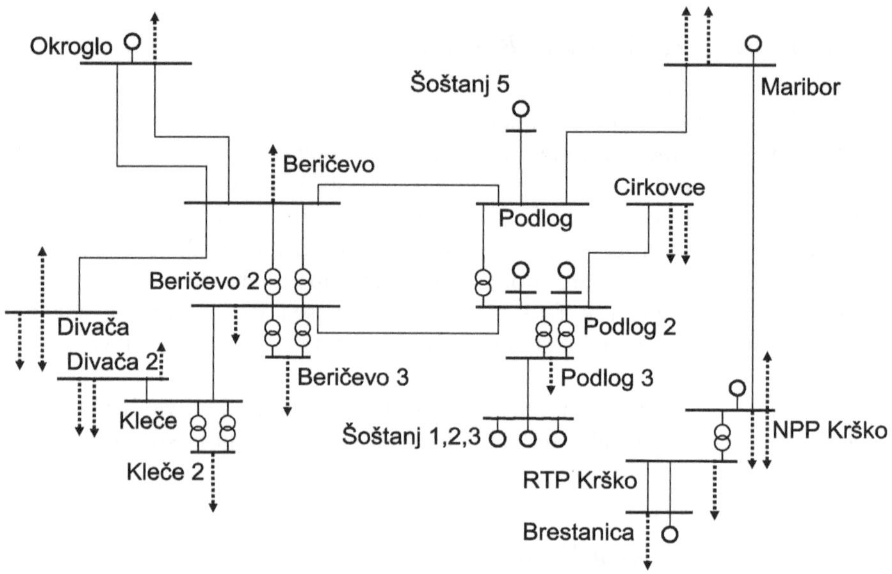

Fig. 15.10 Regional power system configuration

in an island mode of operation as an alternative offsite power source. The power flows through interconnections with neighboring power systems are considered in the loads of the corresponding substations, where those lines are connected.

The representative generating units in the substations connected to neighboring power systems were added in the model to consider power flows from neighboring power systems in substations Divača, Divača 2, Maribor, or adjacent hydro power plants connected to the 110-kV network in substations Maribor, Podlog 2, and Okroglo.

Several interesting power system upgrades were considered from the reliability viewpoint in addition to the basic system evaluation. The basic system configuration is considered as case 1. The new interconnection: single power line and double power line between substations Krško and Beričevo were introduced to the

Table 15.6 Regional power system configurations

No	Regional power system model
1	Basic configuration of the regional power system
2	Regional power system and added new single line between substations Krško and Beričevo
3	Regional power system and added new double line between substations Krško and Beričevo
4	Regional power system and added new nuclear power plant, proportional load increase, and added single line Krško–Beričevo
5	Regional power system and added new nuclear power plant, proportional load increase, and added double line Krško–Beričevo
6	Regional power system and added new nuclear power plant, load increase in Divača, and added single line Krško–Beričevo
7	Regional power system and added new nuclear power plant, load increase in Divača, and added double line Krško–Beričevo

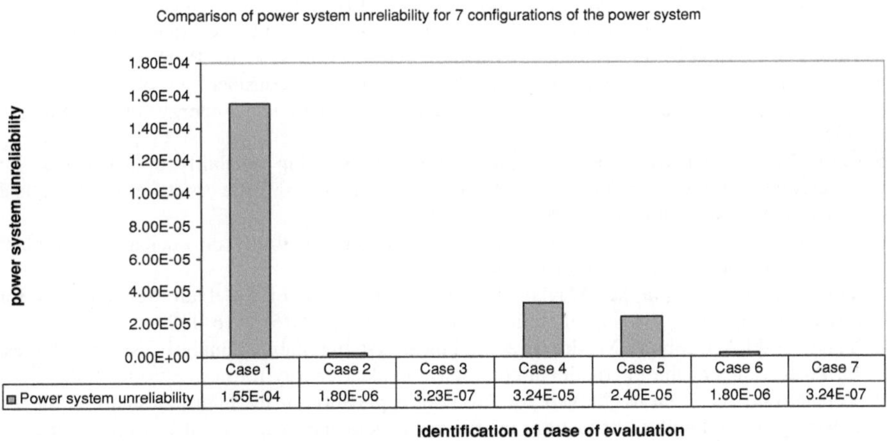

Fig. 15.11 Regional power system results

basic configuration. Those two configurations represent case 2 and case 3. A new nuclear power plant with the parameters equal to the existing nuclear power plant was added to the corresponding substation. The power system configuration with new nuclear power plant is considered in four cases: case 4, case 5, case 6, and case 7, varying two locations of increased loads because of new generation plant and comparing the option with single or with double power line between Krško and Beričevo. The increased power system load because of new large power plant was considered in two ways. One way includes the increase of the complete load only in one substation, i.e., substation Divača, which is a realistic situation corresponding to the export of power to the neighboring power system, where there is a long-term deficiency of power. The other way of increasing the overall load of the power system is the increase of the loads in the power system proportionally to their existing loads.

Table 15.6 shows the summary of the cases of evaluation. Figure 15.11 shows the results of the cases of evaluation.

The results of considered evaluations cases show that adding a new generating unit or adding a new power line between important substations significantly increases the reliability of the power system. Thus, the unreliability of its complement is decreased.

The developed method with the necessary specific modifications is applicable for the estimation of the reliability of other systems of similar topology, such as infrastructure systems for gas and water, computer, transport, and various goods distribution systems.

References

1. Brown RE (2002) Electric power distribution reliability. CRC Press, Boca raton
2. Short TA (2006) Distribution reliability and power quality. Taylor & Francis, Boca Raton
3. Billinton R, Allan R (1996) Reliability evaluation of power systems. Plenum, New York
4. Čepin M (2005) Method for assessing reliability of a network considering probabilistic safety assessment. In: Proceedings of the international conference nuclear energy for new Europe, NSS
5. Čepin M (2006) Development of new method for assessing reliability of a network. In: Proceedings of the eight international conference on probabilistic safety assessment and management, New Orleans, IAPSAM 45/1-45/8
6. Volkanovski A (2008) Impact of offsite power system reliability on nuclear power plant safety. PhD thesis, University of Ljubljana
7. Volkanovski A, Čepin M, Mavko B (2009) Application of the fault tree analysis for assessment of power system reliability. Rel Eng Syst Saf 94(6):1116–1127
8. Volkanovski A, Čepin M, Mavko B (2006) Power system reliability analysis using fault trees. In: Proceedings of the international conference nuclear energy for new Europe, NSS
9. Volkanovski A, Čepin M, Mavko B (2007) An application of the fault tree analysis for the power system reliability estimation. In: Proceedings of the international conference nuclear energy for new Europe, Portorož
10. Volkanovski A, Čepin M, Mavko B (2008) Application of the fault tree analysis for assessment of the power system reliability. In: Proceedings of the ESREL2008 & 17th SRA EUROPE Valencia, ESRA
11. Kančev D, Čauševski A, Čepin M, Volkanovski A (2007) Application of probabilistic safety assessment for macedonian electric power system. In: Proceedings of the international conference nuclear energy for new Europe, NSS
12. Grigg C, Wong P, Albrecht P, Allan R, Bhavaraju M, Billinton R, Chen Q, Fong C, Haddad S, Kuruganty S, Li W, Mukerji R, Patton D, Rau N, Reppen D, Schneider A, Shahidehpour M, Singh C (1999) The IEEE Reliability Test System—1996: a report prepared by the Reliability Test System Task Force of the Application of Probability Methods Subcommittee. IEEE Trans Power Syst 14(3):1010–1020
13. Strategy of transmission system from the year 2007 to 2016 (2007) ELES, Ljubljana
14. Vaurio JK, Tammi P (1995) Modelling the loss and recovery of electric power. Nucl Eng Des 157(1–2):281–293
15. NRC (2005) Reevaluation of station blackout risk at nuclear power plants (NUREG/CR 6890). NRC, Washington
16. NRC (1988) Regulatory guide 1.155: station blackout. NRC, Washington
17. Jeffrey SM (2006) The NRC and Grid Stability, Commissioner, Nuclear Regulatory Commission: ANS executive conference on grid reliability, stability and off-site power. Denver

18. NRC (2006) Grid reliability and the impact on plant risk and the operability of offsite power. Generic letter 2006-02. NRC, Washington
19. NUREG-1784 (2003) Operating experience assessment: Effects of grid events on nuclear power plant performance. NRC, Washington
20. NUMARC 87-00 (1988) Guidelines and technical bases for NUMARC initiatives addressing station blackout at light water reactors. NRC, Washington
21. NUREG-1032 (1988) Evaluation of station blackout accidents at nuclear power plants. NRC, Washington
22. Carreras A, Newman DE, Dobson I, Poole AB (2000) Evidence for self-organized criticality in electric power system blackouts. In: Proceedings of the 33rd annual Hawaii international conference on system sciences
23. Koonce AM, Apostolakis GE, Cook BK (2006) Bulk power grid risk analysis: ranking infrastructure elements according to their risk significance, ESD-WP-2006-19. Engineering Systems Division, working paper series, MIT
24. Carreras BA, Lynch VE, Dobson I, Newman DE (2002) Critical points and transitions in an electric power transmission model for cascading failure blackouts. Chaos 12(4):985–994
25. Apostolakis GE, Lemon DM (2005) Screening methodology for the identification and ranking of infrastructure vulnerabilities due to terrorism. Risk Anal 25(2):361–376
26. Garrick BJ, Hall JE, Kilger M et al (2004) Confronting the risk of terrorism: making the right decisions. Rel Eng Syst Saf 86:129–176
27. Patterson SA, Apostolakis GE (2007) Identification of critical locations across multiple infrastructures for terrorist actions. Rel Eng Syst Saf 92(9):1183–1203
28. Wei-Chang Y (2007) An improved sum-of-disjoint-products technique for the symbolic network reliability analysis with known minimal paths. Rel Eng Syst Saf 92(2):260–268
29. Miki T, Okitsu D, Kushida M, Ogino T (1999) Development of a hybrid type assessment method for power system dynamic reliability. In: Proceedings of the IEEE international conference on systems man and cybernetics IEEE SMC'99, pp 968–973
30. Zio E, Podofillini L, Zille V (2006) A combination of Monte Carlo simulation and cellular automata for computing the availability of complex network systems. Rel Eng Syst Saf 91:181–190
31. Yishan L (2005) Short-term and long-term reliability studies in deregulated power system. PhD thesis, Texas A&M University
32. Yifeng L (2003) Bulk system reliability evaluation in a deregulated power industry. Master's thesis, University of Saskatchewan, Saskatoon, 2003
33. Rajesh UN (2003) Incorporating substation and switching station related outages in composite system reliability evaluation. Master's thesis, University of Saskatchewan, Saskatoon
34. Hua C (2000) Generating system reliability optimization. PhD thesis, University of Saskatchewan, Saskatoon
35. Billinton R, Wangdee W (2006) Delivery point reliability indices of a bulk electric system using sequential Monte Carlo simulation. IEEE Trans Power Delivery 21(1):345–352
36. Haarla L, Pulkkinen U, Koskinen M et al (2008) A method for analysing the reliability of a transmission grid. Rel Eng Syst Saf 93(2):277–287
37. Pottonen L (2005) A method for the probabilistic security analysis of transmission grids. A doctoral dissertation, Helsinki University of Technology
38. Kumamoto H, Henley EJ (1996) Probabilistic risk assessment and management for engineers and scientists. IEEE, New York
39. Roberts NH, Vesely WE, Haasl DF, Goldberg FF (1981) Fault tree handbook: NUREG-0492. NRC, Washington
40. Vesely W, Dugan J, Fragola J et al (2002) Fault tree handbook with aerospace applications. National Aeronautics and Space Administration
41. MIL-HDBK-338B (1998) Department of Defense
42. Mosleh A, Fleming KN (1988) Procedures for treating common cause failures in safety and reliability studies (NUREG/CR-4780). NRC, Washington

43. Čepin M (2005) Analysis of truncation limit in probabilistic safety assessment. Rel Eng Syst Saf 87(3):395–403
44. Vesely WE, Davis TC, Denning RS et al (1983) Measures of risk importance and their applications (NUREG/CR-3385). NRC, Washington
45. Hamzehee H, Hilsmeier T (2008) Use of risk insights in support of USNRC reviews of new reactor applications. PSAM 9, Hong Kong
46. Allan RN, Billinton R (1992) Probabilistic methods applied to electric power systems: are they worth it? Power Eng J 6(3):121–129B
47. IEEE Std 1366 (2003) IEEE guide for electric power distribution reliability indices. IEEE
48. Grainger JJ, Stevenson WD (1994) Power system analysis. McGraw-Hill, New York
49. Ačkovski R (1989) Contribution on methods for planning and development of power systems using Monte Carlo simulation. PhD thesis, Faculty of Electrical Engineering, Skopje
50. Todorovski M (1995) Approximate calculation of power flows thought high voltage network. Graduation work, Faculty of Electrical Engineering, Skopje
51. Aswani D, Badreddine B, Malone M et al (2008) Criteria for evaluating protection from single points of failure for partially expanded fault trees. Rel Eng Syst Saf 93(2):206–216
52. Data development (1985) NUREG/CR 4350/6. NRC, Washington
53. NUREG-5496 (1997) Evaluation of loss of offsite power events at nuclear power plants: 1980–1996. NRC, Washington
54. IEEE Standard 500 (1984) IEEE guide to the collection and presentation of electrical, electronic, sensing component, and mechanical equipment reliability data for nuclear-power generating stations, Appendix D, reliability data for nuclear-power generating stations. IEEE
55. TUD Office and Pörn Consulting (2000) T-book: reliability data of components in Nordic nuclear power plants. Villingby, Sweden, TUD Office and Pörn Consulting
56. IAEA-TECDOC-478 (1988) Component reliability data for use in probabilistic safety assessment. IAEA
57. Čepin M (2010) Application of common cause analysis for assessment of reliability of power systems. In: Proceedings of the PMAPS, IEEE, pp 575–580

Part V
Optimization Methods

Chapter 16
Linear Programming

If we did all the things we are capable of, we would literally astound ourselves

Thomas Edison

16.1 Introduction

Linear programming is an optimization method capable of dealing with an objective function and constraints written as linear inequalities related to objective function and finding the optimal value under specified constraints [1–7].

The term linear programming refers to the linear mathematical relationship among the variables of the objective function or the criterion of the optimization and among the variables of the constraints or restrictions, which apply to the objective function.

The objective function is a measure of the performance of the activity under investigation and can be, for example, the smallest costs, the largest profit, the minimal risk, or the maximal system reliability.

The constraints are restrictions that apply to the parameters of the optimization and are expressed as inequality equations. Feasible solution is the parameter value that satisfies all the constraints. If the objective function does not have any feasible solution, it is called an infeasible function.

The optimal solution corresponds to the optimal value of the objective function, which is either minimal or maximal value, as it is specified by the objective function. Several optimal solutions are theoretically possible. If the solution would be infinitely high or low, it is called an unbounded solution.

16.2 Mathematical Model

The standard mathematical model consists of the objective function and constraints. The objective function can be written in the form of linear equation of parameters [8, 9].

$$\sum_{i=1}^{n} c_i x_i = \min \quad \text{or} \quad \sum_{i=1}^{n} c_i x_i = \max \tag{16.1}$$

M. Čepin, *Assessment of Power System Reliability*,
DOI: 10.1007/978-0-85729-688-7_16,

The constraints are written in the form of linear inequalities.

$$
\begin{aligned}
& a_{11}x_1 + a_{12}x_2 + \cdots + a_{1n}x_n \le b_1 \\
& a_{21}x_1 + a_{22}x_2 + \cdots + a_{2n}x_n \le b_2 \\
& \ldots \\
& a_{m1}x_1 + a_{m2}x_2 + \cdots + a_{mn}x_n \le b_m \\
& x_1 \ge 0 \\
& \ldots \\
& x_n \ge 0
\end{aligned}
\tag{16.2}
$$

A large number (n) of variables is possible. A large number ($n + m$) of inequalities is possible.

16.3 Simple Example

The general process for solving the simplest linear programming exercises is to graph the inequalities, which represent constraints to form an area on the *x, y* plane with possible solutions, which represent the feasibility region. Calculation of the coordinates of the corners of the feasibility region is obtained by finding the intersection points. The simplest procedure of getting the minimal or maximal value of the objective function is to compare the values of the objective function in the intersection points.

The example objective function, for which the maximal value is to be determined, is:

$$z = 2x + 5y \tag{16.3}$$

The following constraints are determined:

$$y - 2x + 2 < 0 \tag{16.4}$$

$$y + 1.2x - 5 < 0 \tag{16.5}$$

$$y - x + 2 > 0 \tag{16.6}$$

Figure 16.1 shows the graphical solution of finding the feasible solutions. Semi-plane areas determined by the inequality constraints are drawn in different textures in the *left variant* of the figure. Their intersection is shown in the *right variant* of the figure and identifies the triangle of feasible solutions.

Table 16.1 shows the solution for simple example. The values for parameters x and y are calculated from intersection of *upper lines*. The objective function is calculated from obtained parameters. The obtained maximal value of the objective function is compared with values of the objective function at other intersections

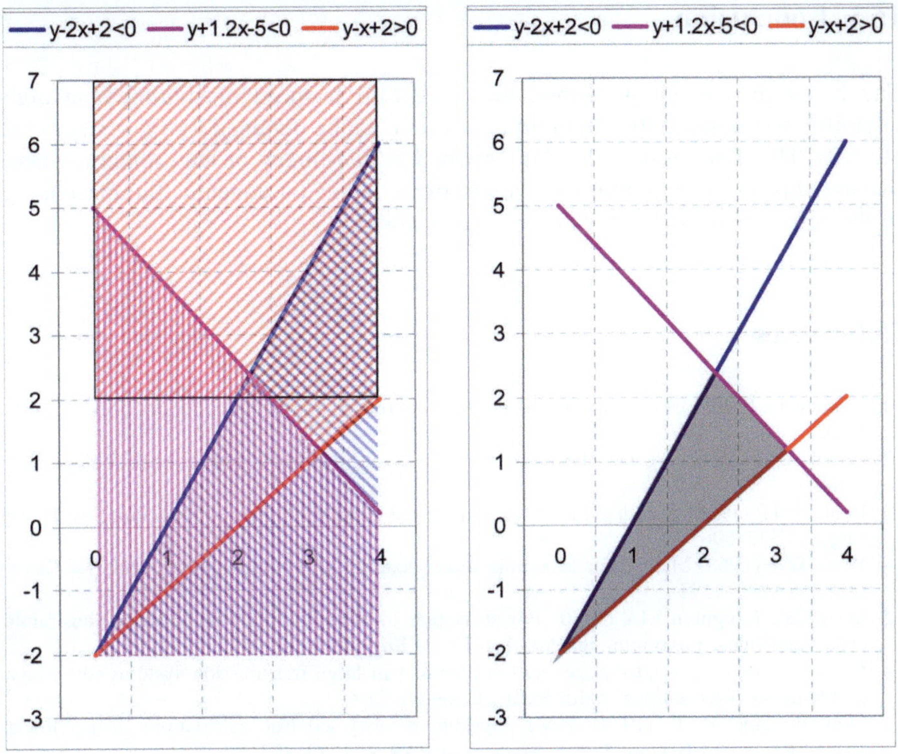

Fig. 16.1 Graphical representation of feasible solutions of simple example

Table 16.1 Solution of simple example

Upper cross-point from the figure	
x	2.1875
y	2.375
$z = 2x + 5y$	16.25

and no better value of the objective function is identified, which confirms the calculated optimum.

16.4 More Complex Problems

For more complex problems with larger number of variables and larger number of constraints, an optimization procedure called simplex procedure is developed [8–12].

16.5 Conclusion

The linear programming method has a very high speed of solution and high reliability in the sense that an optimal solution can be obtained for most situations [13–18]. The main drawback of the method is inaccuracy of the problem, where linearized problem was built from a non-linear one. Consequently, the inaccuracy of the result follows the inaccuracy of the model.

References

1. Momoh JA (2005) Electric power system applications of optimization. Marcel Dekker, New York
2. Wood AJ, Woolenberg BF (1996) Power generation, operation, and control. Wiley, New York
3. Benthall TP (1968) Automatic load scheduling in a multiarea power system. Proc Inst Electr Eng 115:592–596
4. Wells DW (1968) Method for economic secure loading of a power system. Proc Inst Electr Eng 115:1190–1194
5. Shen CM, Laughton MA (1970) Power system load scheduling with security constraints using dual linear programming. Proc Inst Electr Eng 117:2117–2127
6. Merlin A (1972) On optimal generation planning in large transmission systems (the Maya problem). In: Proceedings of 4th PSCC, Grenoble 21-6
7. Stott B, Hobson E (1978) Power system security control calculations using linear programming. IEEE Trans Power Appar Syst PAS 97:1713–1731
8. Vanderbei RJ (2008) Linear programming: foundations and extensions. International series in operations research and management science, vol 114. Springer, New York
9. Schrijver A (1998) Theory of linear and integer programming. Wiley, New York
10. Gärtner B, Matoušek J (2006) Understanding and using linear programming. Springer, Berlin, New York
11. Padberg M (1999) Linear optimization and extensions. Springer, Berlin, New York
12. Dantzig GB, Thapa MN (1997) Linear programming 1: introduction. Springer, New York
13. Dantzig GB, Thapa MN (2003) Linear programming 2: theory and extensions. Springer, New York
14. Alevras D, Padberg MW (2001) Linear optimization and extensions: problems and solutions. Springer, Berlin, New York
15. Ferguson TS (1998) Linear programming: a concise introduction. http://www.usna.edu/Users/weapsys/avramov/Compressed%20sensing%20tutorial/LP.pdf. Accessed 18 Dec 2010
16. Strayer JK (1989) Linear programming and its applications. Springer, New York
17. Press WH, Teukolsky SA, Vetterling WT, Flannery BP (2007) Numerical recipes: the art of scientific computing. Cambridge University Press, Cambridge, UK
18. Zhu J (2009) Optimization of power system operation. Wiley, Chichester

Chapter 17
Dynamic Programming

No human investigation can be called real science if it cannot be demonstrated mathematically

Leonardo da Vinci

17.1 Introduction

Dynamic programming is an optimization method that transforms a complex problem into a sequence of simpler problems [1–5]. A sequence of simpler problems can be dealt with a variety of optimization methods that can be employed to solve particular aspects of a more general formulation. Dynamic programming can be top-down or bottom-up oriented [6–10].

17.2 Method

A dynamic programming is a method that is usually based on a recurrent formula and initial states. A subsolution of the problem is constructed from previously found ones. Dynamic programming solutions have a polynomial complexity that assures a much faster running time than the other techniques, such as backtracking or brute-force technique.

Three most important characteristics of dynamic programming problems are the following:

- Multiple stages, which are solved sequentially one stage at a time.
- States, which reflect the information required to assess the consequences that the current decision has upon future actions.
- Recursive optimization, which builds to a solution of the overall N-stage problem by first solving a one-stage problem and sequentially including one stage at a time and solving one-stage problems until the overall optimum has been found.

Mathematical model includes a multistage decision process where the cost c_n for a particular stage n is:

DOI: 10.1007/978-0-85729-688-7_17,

$$c_n = f_n(d_n,\ s_n) \tag{17.1}$$

where d_n is a permissible decision that may be chosen from the set D_n and s_n is the state of the process with n stages to go.

Let us assume that there are a total of N stages in the process and we continue to think of n as the number of stages remaining in the process. The next state of the process depends entirely on the current state of the process and the current decision taken.

$$s_{n-1} = t_n(d_n,\ s_n) \tag{17.2}$$

where d_n is the decision chosen for the current stage and state.

The objective is to maximize the sum of the cost functions over all stages of the decision process and the constraints on this optimization are that the decision chosen for each stage belongs to some set D_n of permissible decisions and that the transitions from state to state be governed by previous equation. As we are in state s_n with n stages to go, the optimization problem is to choose the decision variables $d_n, d_{n-1}, \ldots, d_0$ to solve the following problems:

$$c_n(s_n) = \text{Min}\Big[f_n(d_n,\ s_n) + f_{n-1}(d_{n-1},\ s_{n-1}) + \cdots + f_0(d_0,\ s_0)\Big] \tag{17.3}$$

subject to:

$$s_{m-1} = t_m(d_m,\ s_m); \qquad m = 1, 2, \ldots, n \tag{17.4}$$

$$d_m \in D_m; \qquad m = 1, 2, \ldots, n \tag{17.5}$$

The function $c_n(s_n)$ is the optimal value function and it represents the minimal cost possible over the n stages. The function $c_n(s_n)$ is the optimal value of all subsequent decisions, given that we are in state s_n with n stages to go.

Function $f_n(d_n,\ s_n)$ involves only the decision variable d_n and not the decision variables $d_{n-1}, \ldots, d_0$, so we could first maximize over this latter group for every possible d_n and then choose d_n so as to maximize the entire expression. Therefore, we can rewrite previous equation.

$$c_n(s_n) = \text{Min}\Big[f_n(d_n,\ s_n) + \text{Min}\{f_{n-1}(d_{n-1},\ s_{n-1}) + \cdots + f_0(d_0,\ s_0)\}\Big] \tag{17.6}$$

subject to:

$$s_{n-1} = t_n(d_n,\ s_n); \quad s_{m-1} = t_m(d_m,\ s_m), \quad m = 1, 2, \ldots, n-1 \tag{17.7}$$

$$d_n \in D_n, d_m \in D_m; \qquad m = 1, 2, \ldots, n-1 \tag{17.8}$$

The second part of the previous equation is simply the optimal value function for the $n - 1$ stage dynamic programming problem; so we can, therefore, rewrite the equation into recursive form or in a form of a formal statement of the principle of optimality:

$$c_n(s_n) = \text{Min } [f_n(d_n, s_n) + c_{n-1}(s_{n-1})] \tag{17.9}$$

subject to:

$$s_{n-1} = t_n(d_n, s_n) \tag{17.10}$$

$$d_n \in D_n \tag{17.11}$$

To emphasize the optimization over d_n, equation can be rewritten in the other form of a formal statement of the principle of optimality.

$$c_n(s_n) = \text{Min}[f_n(d_n, s_n) + c_{n-1}[t_n(d_n, s_n)]] \tag{17.12}$$

subject to:

$$d_n \in D_n \tag{17.13}$$

The starting value has to be determined because of the recursive algorithm.

$$c_0(s_0) = \text{Min}[f_0(d_0, s_0)] \tag{17.14}$$

subject to:

$$d_0 \in D_0 \tag{17.15}$$

The example applications are presented in older [5, 11, 12] and more recent references [13–15].

References

1. Bellman R (2003) Dynamic programming. Denardo
2. Bellman R (1957) Dynamic programming. Princeton University Press, Princeton, NJ
3. Momoh JA (2005) Electric power system applications of optimization. Marcel Dekker, New York
4. Wood AJ, Woolenberg BF (1996) Power generation operation and control. Wiley, New York
5. Bradley S, Hax A, Magnanti T (1977) Applied mathematical programming. Addison-Wesley, Reading, MA
6. Dasgupta S, Papadimitriou CH, Vazirani UV (2008) Algorithms. McGraw-Hill, Boston, MA
7. Sniedovich M (2010) Dynamic programming: foundations and principles. Taylor & Francis,
8. Bertsekas DP (2007) Dynamic programming and optimal control. Athena Scientific, Belmont, MA
9. Borkar VS (2000) Average cost dynamic programming equations for controlled Markov chains with partial observations. SIAM J Contr Opt 39:673–681
10. Zhu J (2009) Optimization of power system operation. Wiley, New York
11. Howard RA (1960) Dynamic programming and Markov processes. Wiley, New York
12. Hadley G (1962) Nonlinear and dynamic programming. Addison-Wesley, London
13. Zietz J (2004) Dynamic programming: an introduction by example. http://frank.mtsu.edu/~berc/working/Zietz-DP-1.pdf. Accessed 14 Jan 2011
14. Introduction to dynamic programming http://www.webmath.iitkgp.ernet.in/breadth/ma23011/lectnotes07/Dynamic_Programming-4.pdf. Accessed 14 Jan 2011
15. Chinneck JV (2006) Practical optimization: a gentle introduction. http://www.sce.carleton.ca/faculty/chinneck/po.html. Accessed 14 Jan 2011

Chapter 18
Genetic Algorithm

The true delight is in the finding out rather than in the knowing
Isaac Asimov

18.1 Introduction

Natural selection has provided the nature with a various types of species. The organisms, which are the representatives of species, evolve and adapt to their environment. Better-adapted organisms reproduce and survive through the generations. They transfer their genetic material that made them successful to their offspring. Some species are faced to extinction. Some other species evolve to maintain the balance in nature [1, 2].

Based on the concept of natural evolution, an optimization tool for engineering problems was developed [1–3]. The idea was to evolve a population of candidate solutions to a given problem, using operators inspired by natural genetic variation and natural selection.

Similar methods that simulate the processes from nature are simulated annealing [4], particle swarm optimization [5], and ant colony optimization [5]. They represent processes in nature that are remarkably successful at optimizing natural phenomena [6–22].

18.2 Definition and Use

Genetic algorithm is a probabilistic search method founded on the principle of natural selection and genetic recombination [23]. Genetic algorithm represents a powerful method that efficiently uses historical information to evaluate new search points with expected better performance. It is applicable to linear and to nonlinear problems with many local extrema.

Genetic algorithm was introduced by Holland [24]. De Jong [7] showed the usefulness of the genetic algorithms for function optimization and made the first concerted effort to find optimized genetic algorithm parameters. Goldberg solved a

M. Čepin, *Assessment of Power System Reliability*,
DOI: 10.1007/978-0-85729-688-7_18, © Springer-Verlag London Limited 2011

problem involving the control of gas pipeline transmission for his dissertation [3]. During the past several years, genetic algorithms have been applied to diverse types of problems ranging from scheduling to computational biology.

18.3 Genetic Algorithm Representations

Two most used types of genetic algorithms representations are binary representation and the real-valued representation [8–11]; therefore, the terminology binary genetic algorithms and real genetic algorithms is often in use. Both algorithms follow the same menu of modeling natural selection and genetic recombination. One represents variables as an encoded binary string and works with the binary strings to explore the search space, whereas the other works with the real-valued variables themselves to explore the search space. Binary genetic algorithm was first introduced and mainly used but in recent time, the real genetic algorithms are more recommended.

The genetic algorithm begins by defining the type of representation, the fitness function, and the variables. After which both genetic algorithm representations start with the initial population form N chromosomes of random values. Each of the chromosomes represents a possible solution for the fitness function. The main difference between the algorithms is that the binary genetic algorithm is representing each of the variables as binary string, whereas the real genetic algorithm is using the real value of the variables themselves. For example, two chromosome examples of the genetic algorithm types described above are the following:

- Binary-coded chromosome: [1101101, 0010011, . . . , 0101100]
- Real-coded chromosome: [109, 19, . . . , 44]

After the population is created, the fitness of each chromosome is evaluated using the fitness function, which is the measurement for the quality of each solution. Better solutions get higher scores, with the global optima having the highest value. Chromosomes are selected to reproduce using crossover and mutation operators according to their fitness. Figure 18.1 shows two flow charts representing the binary genetic algorithms and the real genetic algorithms.

18.4 Genetic Algorithm Structure

There are various types of genetic algorithms not only from perspective how variables are represented (e.g., binary and continuous) but also by the replacement policy and the type of genetic algorithm operators. Simple genetic algorithm is the simplest type of genetic algorithm. Pseudocode of simple genetic algorithm is shown in Table 18.1.

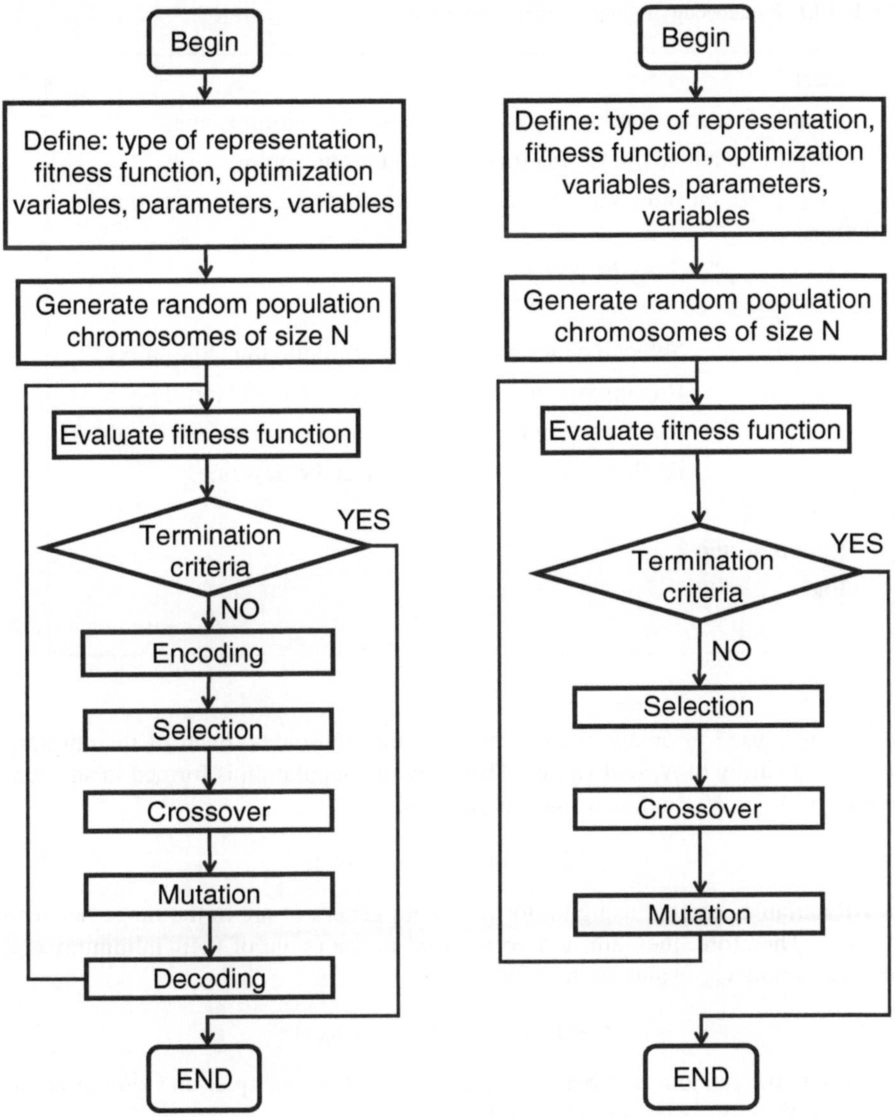

Fig. 18.1 Genetic algorithm flow charts

18.4.1 Initial Population

How to initialize the genetic algorithm population is a key decision. Initializing the population can be done by randomly generating a number of chromosomes, each one representing a different solution to the problem. Usually, uniform random

Table 18.1 Pseudocode of simple genetic algorithm

```
Start
(step 1) Generate random initial population of N chromosomes
(step 2) Calculate the fitness f(x) for each chromosome
(step 3) If the number of iteration is equal to the maximum number
defined
                Stop the procedure
        Else
                Select the chromosomes for reproduction (N/2 pairs)
                Run crossover operator (create new population)
                Run mutation operator (introduce new random variables)
                Replace the old population with the new one
        End
Return to step 2
End
```

generator is used to create population of N_p chromosomes. Each of the chromosomes is an array of N_v real values. Therefore the population is formed in shape of matrix of $N_p \times N_v$ random values as follows:

$$\text{POP} = \text{random}\left(N_p \times N_v\right) \tag{18.1}$$

All variables created using uniform random generator are in the range between 0 and 1. Therefore, they are non-normalized in the range of their minimum $x_{\min}$ and maximum $x_{\max}$ value as follows:

$$x = x_{\min} + r \times (x_{\max} - x_{\min}) \tag{18.2}$$

where r is the random number between 0 and 1. For example, one simple population of $N_p \times N_v$ is presented as follows:

$$\text{POP} = \begin{bmatrix} x_1^1 & x_2^1 & \cdots & x_{N_v}^1 \\ x_1^2 & x_2^2 & \cdots & x_{N_v}^2 \\ \vdots & \vdots & \ddots & \vdots \\ x_1^{N_p} & x_2^{N_p} & \cdots & x_{N_v}^{N_p} \end{bmatrix} \tag{18.3}$$

Each row represents one chromosome (i.e., one solution) and $x_1, x_2, \ldots, x_v$ are the variable values.

18.4.2 Fitness Function

The fitness function is by definition the measure of how well a candidate solution solves the problem. It is important to distinguish between fitness function and objective function that are used in genetic algorithms. The *fitness function*, which may be defined as the sum of all objective function value and the penalty for constraint violation, can be calculated for each chromosome.

$$\mathrm{FIT}_t = \frac{1}{\sum_j f_i(x_{ij}) + \sum_k \mathrm{FP}_k} \tag{18.4}$$

where

$$\mathrm{FP}_k = \delta_k \times \mathrm{VIOL}_k \tag{18.5}$$

where VIOL_k is the violation of constraint k and δ_k is the penalty multiplayer of constraint k. For better understanding of previous equations, examples of objective functions are given.

$$f_1(x_1, x_2) = x_1^2 + x_2^2 - 4x_1x_2 + 10 \tag{18.6}$$

$$f_2(x_1, x_2) = x_1^2 - x_2^2 + 2x_1x_2 + 15 \tag{18.7}$$

Subject to the constraints:

$$0 < x_1 < 10 \quad 0 < x_2 < 15 \quad x_1 + x_2 = 20 \tag{18.8}$$

therefore:

$$\mathrm{OBJ}_{\mathrm{fun}} = f_1(x_1, x_2) + f_2(x_1, x_2) \tag{18.9}$$

Equations 18.1–18.4 can also be represented in the inverse form which is important due the process of minimization or maximization. In case, if minimization of the objective function is required, after fitness for each chromosome is calculated using Eqs. 18.1–18.4, the chromosomes with greater fitness values are assigned with greater probability for reproduction. But when maximization is required, the chromosomes with lower fitness value are assigned with greater probability for reproduction. In case when fitness function from mentioned equations is presented in its inverse form, the explanation that was given above is in opposite order.

18.4.3 Genetic Operators

Selection, recombination, and mutation are the operators used to ensure reproduction in the genetic algorithms. Therefore, it is important that they are properly used according to the problem specified.

Table 18.2 The selection step using fitness proportionate selection

Number of chromosome	Fitness (FIT_i)	Probability (P_i)	Cumulative probability $\left(\sum_{i=1}^{n} P_i\right)$	Selected chromosomes
C1	112	112/350 = 0.320	0.320	C2
C2	104	104/350=0.297	0.617	C1
C3	40	40/350=0.114	0.731	C3
C4	34	34/350=0.097	0.828	C1
C5	33	33/350=0.094	0.922	C6
C6	27	27/350=0.077	1	C2
Total	350	1		

18.4.3.1 Selection

The selection process follows the evaluation of the fitness function. The selection is a procedure in which chromosomes are selected for reproduction. The purpose of selection is to emphasize the fitter individuals in the population in hope that their offspring will in turn have even better fitness. Selection has to be balanced with variation from crossover and mutation. Too strong selection pressure means that only highly fit individuals will take over the population, reducing the diversity needed for further change and progress. Too weak selection pressure results in too slow evolution. Numerous selection schemes have been proposed in the genetic algorithm literature. Some of the most important are presented.

Fitness Proportionate Selection

Fitness proportionate selection is also known as roulette wheel selection. The method is conceptually equivalent to giving each chromosome a slice of a circular roulette wheel equal in area to the individual fitness. This means that the probability of an individual to be chosen for reproduction is proportional to its fitness. Those with higher fitness are selected more often than individuals with lower fitness. The probability for selection of an individual whose fitness is FIT_i is given by the ratio of the individual fitness to the total fitness of all the individuals in that population.

$$P_i = \frac{\text{FIT}_i}{\sum_j \text{FIT}_j} \tag{18.10}$$

The cumulative probabilities calculated as are used in selecting the chromosomes [10]. A uniformly distributed random number between zero and one is generated. Starting at the top of the list, the first chromosome with a cumulative probability value that is greater than the random number is selected for reproduction. The procedure stops when the number of chromosomes selected is equal with the number of chromosomes from the population. Table 18.2 shows a sample population of six individuals, their assumed fitness, the probability, the cumulative probability, and the selected chromosomes.

When fitness proportionate selection is applied to all population, $N_p/2$ pairs of chromosomes (i.e., parents) are chosen for reproduction. By default, chromosomes with higher selection probability will be present more often in the parent population, as it is shown in Table 18.2.

At the beginning of the search, the differences in fitness in the population are usually high (Table 18.2). This means that small number of chromosomes is much fitter than the others. Under fitness proportional selection, they and their offspring multiply quickly in the population, preventing the genetic algorithm from doing any further exploration. This is known as premature convergence. In other words, fitness proportionate selection early in the search puts too much pressure on exploitation of highly fit chromosomes, neglecting the exploration in other regions of the search space. When all individuals in the population are very similar, there are no real fitness differences for selection to exploit, and evolution nearly stops.

Rank Selection

Rank selection is another roulette wheel type of selection. It designed to prevent premature convergence as an alternative of fitness proportionate selection. This method is based on ranking of the population. The expected selection probability value of each chromosome depends on its rank rather than on its absolute fitness. Because absolute differences in fitness are irrelevant, there is no need to scale fitness. Exclusion of absolute fitness information can have advantages and disadvantages. Exclusion of absolute fitness information can be an advantage, because the use of absolute fitness can lead to convergence problems. Exclusion of absolute fitness information can be a disadvantage, because in some cases, it might be important to know that one individual is far fitter than its nearest competitor. Ranking avoids giving the far largest share of offspring to a small group of highly fit chromosomes. Thus, reducing the selection pressure when the fitness differences are high and keeping up selection pressure when the fitness differences are low. In other words, the ratio of expected values of chromosomes with ranks n and $n + 1$ will be the same whether their absolute fitness differences are high or low.

The steps of the simplest rank selection method are the following: each chromosome in the population is ranked in increasing order from 1 to N_p, starting with the chromosome with highest fitness to the chromosome with lowest fitness. The probability for selection is calculated as follows:

$$P_i = \frac{N_p + 1 - n}{\sum_{n=1}^{N_p} n} \tag{18.11}$$

For example, the probability for selection of a chromosome with rank of 1 from population of six individuals is calculates as follows:

$$P_1 = \frac{6 + 1 - 1}{1 + 2 + 3 + 4 + 5 + 6} = \frac{6}{21} = 0.286 \tag{18.12}$$

Table 18.3 The selection step using rank selection

Number of chromosomes	Fitness (FIT_i)	Rank	Probability (P_i)	Cumulative probability $\left(\sum_{i=1}^{n} P_i\right)$	Selected chromosomes
C1	112	1	0.286	0.286	C2
C2	104	2	0.238	0.524	C1
C3	40	3	0.190	0.714	C3
C4	34	4	0.143	0.857	C1
C5	33	5	0.095	0.952	C6
C6	27	6	0.048	1	C4
Total	350		1		

Same as at fitness proportionate selection, the cumulative probabilities calculated as are used in selecting the chromosomes. A uniformly distributed random number between zero and one is generated. Starting at the top of the list, the first chromosome with a cumulative probability value that is greater than the random number is selected for reproduction. The procedure stops when number of the chromosomes selected is equal with the number of chromosomes from the population. Table 18.3 shows a sample population of six individuals, their assumed fitness, the probability, the cumulative probability and the selected chromosomes using rank selection.

Tournament Selection

Another approach that closely mimics mating competition in nature is the tournament selection method. It is similar to the rank selection in terms of selection pressure, but it is computationally more efficient and more suitable for parallel implementation. Main difference is that the rank selection requires sorting the entire population by rank, which is a potentially time-consuming procedure. Tournament selection randomly picks a small subset of chromosomes (two or three) from the parent population, and the fittest chromosome in this subset becomes a parent. The tournament repeats for every parent needed [12]. Tournament selection works best for larger population sizes because sorting becomes time consuming for large populations.

18.4.3.2 Recombination

Crossover is a recombination technique used for creation of one or more offspring form the parents selected by means of exchange of the genetic material. That is, two parents usually create two offspring. The following techniques are mostly used in the literature [8–12]:

- Single-point crossover
- Two-point crossover
- Uniform crossover

Single-point crossover is the simplest one to code and use. A single crossover position is selected at random and the parts of two parents after the crossover position are exchanged to form two offspring. For example purposes, consider the two parents to be:

$$\text{Parent}_1 = [x_{m1}, x_{m2}, x_{m3}, x_{m4}, x_{m5}, x_{m6}, \ldots, x_{mNv}] \tag{18.13}$$

$$\text{Parent}_2 = [x_{d1}, x_{d2}, x_{d3}, x_{d4}, x_{d5}, x_{d6}, \ldots, x_{dNv}] \tag{18.14}$$

After the crossover point is randomly selected and the variables in between exchanged, the following offspring are created:

$$\text{Offspring}_1 = [x_{m1}, x_{m2}, _\uparrow x_{d3}, x_{d4}, x_{d5}, x_{d6}, \ldots, x_{dNv}] \tag{18.15}$$

$$\text{Offspring}_2 = [x_{d1}, x_{d2}, _\uparrow x_{m3}, x_{m4}, x_{m5}, x_{m6}, \ldots, x_{mNv}] \tag{18.16}$$

At two-point crossover, two positions are selected randomly and the variables between them are exchanged:

$$\text{Offspring}_1 = [x_{m1}, x_{m2}, _\uparrow x_{d3}, x_{d4}, x_{d5}, _\uparrow x_{m6}, \ldots, x_{mNv}] \tag{18.17}$$

$$\text{Offspring}_2 = [x_{d1}, x_{d2}, _\uparrow x_{m3}, x_{m4}, x_{m5}, _\uparrow x_{d6}, \ldots, x_{dNv}] \tag{18.18}$$

In special case, more than two points are selected, choosing which of the two parents contributes its variable at each position. Thus, one goes down the line of the chromosomes and at each variable randomly chooses whether to swap information between the two parents. This method is called uniform crossover:

$$\text{Offspring}_1 = [x_{m1}, x_{d2}, x_{d3}, x_{m4}, x_{d5}, x_{m6}, \ldots, x_{dNv}] \tag{18.19}$$

$$\text{Offspring}_2 = [x_{d1}, x_{m2}, x_{m3}, x_{d4}, x_{m5}, x_{d6}, \ldots, x_{mNv}] \tag{18.20}$$

The problem with these point crossover methods for real representation is that no new information is introduced: Each real value that was randomly initiated in the initial population is propagated to the next generation, only in different combinations. Although this strategy worked fine for binary representation, there is now a continuum of values, and in this continuum, we are merely interchanging two data points. These approaches totally rely on mutation to introduce new genetic material.

To fix this problem, the blending method is introduced by combining variable values from the two parents into new variable values in the offspring [8]:

$$x_{1\text{new}} = x_{mn} - r \times (x_{mn} - x_{dn}) \tag{18.21}$$

$$x_{2\text{new}} = x_{dn} + r \times (x_{mn} - x_{dn}) \tag{18.22}$$

where r is the random variable between 0 and 1, and x_{mn} and x_{dn} are the nth variables in the mother and the father chromosome, respectively.

The variables can be blended by using the same r for each variable or by choosing different r for each variable. These blending methods effectively combine the information from the two parents and choose values of the variables between the values bracketed by the parents.

18.4.3.3 Mutation

Mutation is an operator that allows reappearing of the genetic material, that otherwise might be lost from the population using crossover alone. Mutation is a mechanism that ensures that the algorithm is not stuck in local minima. The number of mutations is determined by choice and is often not more than several percents from the number of variables in the population:

$$N_m = r_m \times \left(N_p \times N_v\right) \tag{18.23}$$

where r_m is the mutation rate.

Random numbers are chosen to select the row and columns of the variables to be mutated. A chosen variable is replaced by a new random variable.

It is highly recommended that the mutation rate is not kept constant during the generations when solving large-scale optimization problems. It is obvious that large number of mutations is helpful during early generations to improve the genetic algorithms exploration capability. This is not the case at the end of the search, when large number of mutation can destruct good formed genetic material. Therefore, small mutation rate is needed at the end of the generations to fine tune the population. That is the reason why dozen of dynamic methods are proposed in the literature, including linear, step linear, and Gauss. For example, when linear mutation is applied, the mutation rate is decreasing gradually from a specific maximum value $r_m^{\max}$ to specific minimum value $r_m^{\min}$ with the increase of the number of iterations [14]:

$$r_m = r_m^{\max}\left(1 - \frac{\text{iter}}{\text{iter}_{\max}}\right) + r_m^{\min} \tag{18.24}$$

where $\text{iter}_{\max}$ is the maximum number of generations in the genetic algorithm.

18.4.4 Replacement Policy

After the reproduction, the replacement step replaces the parent population with the new offspring obtained after mutation. The simplest replacement step is the one that is replacing all the chromosomes from the old population with the new ones. After replacement, the new population is used for the next generation until the maximum number of iterations is reached or some other terminating condition is satisfied. The genetic algorithm that is using this simple method is known as simple genetic algorithm or classical genetic algorithm. The individual with the best fitness after the last generation is returned as the solution for the problem. The simplicity of the algorithm makes classical genetic algorithms suitable for application to a broad range of problems.

Because not all chromosomes in the offspring population are better than their parents, a preselection can be applied to improve the classical genetic algorithm features [1]. In preselection, not all generated offspring are chosen for the new population. Only the offsprings with higher fitness than their parents replace their parents in the next generation.

Crowding is another method that is a generalization of preselection where offspring replace similar chromosomes from the previous population. In crowding, selection and reproduction are the same as in the classical genetic algorithm, but replacement is different. It is assumed that two parents are selected to produce two offspring. Two members of the parent population are identified to make room for these offspring. The policy of replacing a member of the present generation by an offspring is carried out as follows:

- Group of individuals is selected randomly from the population (*cw*—the crowding factor and defines the size of the group).
- The bit strings in the offspring chromosome are compared with those of the individuals in the group using the Hamming distance, which measures the number of differing bits, as a measure of similarity.
- The group member that is most similar to the offspring is replaced by the offspring.

Those crowding steps are suitable for binary representation only. For real representation, the only difference is in the middle step, where Euclidian distance is used for measuring of the similarity between the chromosomes.

Not always all of the chromosomes from the population are used for reproduction. In steady-state genetic algorithms, only a subset of the chromosomes in the population, called the generation gap, are replaced in every generation. The parent population and the number of offspring created in every generation are calculated as follows:

$$N_k = N_p \times k \tag{18.25}$$

where k is the generation gap and is a number between 0 and 1.

18.5 Advantages and Disadvantages

Genetic algorithm is a method that provides a robust search in complex spaces. The number of applications of genetic algorithm is growing rapidly because the algorithm is simple and powerful in the search for improvement. Some of the main advantages over the classical optimization algorithms are the following [8].

- The genetic operators are governed by probability and the methodology of using them is problem independent.
- The method and the programs implementing them can be easily applied to many problems with only minor adjustments.
- Solutions are represented with a continuous or discrete variables that allow easy manipulation and fast execution of the genetic operators.
- It is possible to work with several solutions at the same time and improve them by recombining their good features.
- Different areas of the search space are explored while the good solutions guide the algorithm to near optimal solution.
- Escape from local optima is easy.
- The use of an objective function to determine the quality of a solution is the only information to guide the search.
- No derivatives or auxiliary information is needed.
- It is possible to deal with a large number of variables.
- It is well suited for parallel computers.
- The list of optimum variables is provided not just a single solution.
- Genetic algorithm works well with numerically generated data, experimental data, or analytical functions.

Despite some of the very good advantages that genetic algorithm has, there are also some disadvantages over the conventional optimization algorithms. For instance, the traditional methods have the advantage to find quickly the solution of a well-behaved convex analytical function of only a few variables. For such cases, some of the conventional methods outperform the genetic algorithm, quickly finding the minimum while the genetic algorithm is still analyzing the costs of the initial population. For these problems, we should use the already well-known and proven methods that find the solution very quickly. In other words, for problems that are not too difficult, other methods may find the solution faster than the genetic algorithm. However, many realistic problems do not fall into this category. Some other disadvantages of the genetic algorithms are the following.

- Genetic algorithm can be very slow.
- Genetic algorithm cannot always find the exact solution but it always finds the best solution.
- Genetic algorithm shows a very fast initial convergence followed by progressive slower improvements.
- Convergence is difficult in presence of a lot of noise.

18.6 Final Comments

Genetic algorithms are heuristic algorithms with application in many fields in modern science and technology. Various new types of genetic algorithms operators have been applied constantly improving their features after they were developed. The first genetic algorithm application on power systems started in the beginning of the year 1990. Since then, genetic algorithms have been applied not only to pure optimization problems in power systems but also to many other problems in different fields of interest [15–22].

References

1. Cedeno W (1995) The multi-niche crowding genetic algorithm: analysis and application. Doctoral dissertation, University of California
2. Dasgupta D, Michalewicz Z (1997) Evolutionary algorithms in engineering applications. Springer, New York
3. Goldberg DE (1989) Genetic algorithms in search: optimization and machine learning. Addison-Wesley, Reading, Mass
4. Kirkpatrick S, Gelatt CD, Vecchi MP (1983) Optimization by simulated annealing. Science 220(4598):671–680
5. Parsopoulos KE, Vrahatis MN (2002) Recent approaches to global optimization problems through Particle Swarm optimization. Nat Comput 1(2):235–306
6. Schwefel H (1995) Evolution and optimum seeking. Wiley, New York
7. De Jong KA (1975) An analysis of the behavior of a class of genetic adaptive systems. Doctoral dissertation, University of Michigan, Ann Arbor
8. Haupt RL, Haupt SE (2003) Practical genetic algorithms. Wiley, New York
9. Lee KY, El-Sharkawi MA (2008) Modern heuristic optimization techniques: theory and applications to power systems. Wiley, New York
10. Rothlauf F (2006) Representations for genetic and evolutionary algorithms. Springer, Berlin
11. Eiben AE, Smith JE (2003) Introduction to evolutionary computing. Springer, New York
12. Melanie M (1998) An introduction to genetic algorithms. MIT, Cambridge
13. Fogel DB (2006) Evolutionary computing: toward a new philosophy of machine intelligence. Wiley, New York
14. Kumar S, Naresh R (2007) Efficient real code genetic algorithm to solve the non-convex hydrothermal scheduling problem. Electr Power Energy Syst 29:738–747
15. Volkanovski A, Mavko B, Boševski T et al (2008) Genetic algorithm optimization of the maintenance scheduling of generating units in a power system. Rel Eng Syst Saf 93:779–789
16. King TD, El-Hawary ME, El-Hawary F (1995) Optimal environmental dispatching of electric power systems via an improved Hopfield neural network model. IEEE Trans Power Syst 10(3):1559–1565
17. Simopoulos DN, Kavatza SD, Vournas CD (2007) An enhanced peak shaving method for short term hydrothermal scheduling. Energy Convers Manage 48:3018–3024
18. Basu M (2008) Dynamic economic emission dispatch using nondominated sorting genetic algorithm-II. Electr Power Energy Syst 30:140–149
19. Liang RH, Liao JH (2007) A fuzzy-optimization approach for generation scheduling with wind and solar energy systems. IEEE Trans Power Syst 22(4):1665–1674
20. Bharathi R, Kumar MJ, Sunitha D et al (2007) Optimization of combined economic and emission dispatch problem: a comparative study. IEEE Power Eng Conf 134–139
21. Crossley W, Williams EA (1997) A study of adaptive penalty functions for constrained genetic algorithm. In: AIAA 35th aerospace sciences meeting and exhibit, pp 83–97
22. Zhang PX, Zhao B, Cao YJ et al (2004) A novel multi-objective genetic algorithm for economic power dispatch. IEEE Universities Power Eng Conf 422–426
23. Dorigo M, Maria G (1997) Ant colony system: a cooperative learning approach to the traveling salesman problem. IEEE Trans Evol Comput 1(1):53–66
24. Holland JH (1975) Adaptation in natural and artificial systems. University of Michigan Press, Ann Arbor

Chapter 19
Simulated Annealing

Life is not holding a good hand. Life is playing a poor hand well

Danish Proverb

19.1 Introduction

Simulated annealing as an optimization technique attracted significant attention as suitable for solving large-scale optimization problems. The main advantage of simulated annealing is that it can be applied to large problems regardless of the conditions of differentiability, continuity, and convexity that are normally required in conventional optimization methods [1, 2].

Simulated annealing was originally proposed by Metropolis in the early 1950s [3]. The concept is based on the manner in which liquids freeze or metals recrystallize in the process of annealing. In an annealing process, a melt, initially at high temperature, is slowly cooled so that the system at any time is approximately in thermodynamic equilibrium. As cooling proceeds, the system becomes more ordered and approaches a frozen ground state at $T = 0$. Therefore, the process can be thought of as an adiabatic approach to the lowest energy state. If the initial temperature of the system is too low or cooling is done insufficiently slowly, the system may become quenched forming defects or freezing out in metastable states.

The original Metropolis scheme is such that the algorithm generates a sequence of states of a solid, where each state S_i of the solid is with energy E_i and temperature T. Next state S_j is generated by transition mechanism that consists of a small perturbation with respect to the original state, obtained by moving one of the particles of a solid chosen by the Monte Carlo method [2, 3]. Let the energy of the resulting state, which is also found probabilistically, be E_j. If the difference $E_i - E_j$ is less than equal to zero, the new state S_j is accepted. Otherwise, in a case the difference is greater than zero; the new state is accepted with probability:

$$p = e^{\frac{E_i - E_j}{k_b T}} \tag{19.1}$$

where K_b is the Boltzmann constant.

M. Čepin, *Assessment of Power System Reliability*,
DOI: 10.1007/978-0-85729-688-7_19, © Springer-Verlag London Limited 2011

The generalization of this Monte Carlo approach by analogy is applied in combinatorial problems [4, 5]. The current state of the thermodynamic system is analogous to the current solution to the combinatorial problem, the energy equation for the thermodynamic system is analogous to the objective function, and ground state is analogous to the global minimum. The major difficulty in implementation of the algorithm is that there is no obvious analogy for the temperature T with respect to a free parameter in the combinatorial problem. Furthermore, avoidance of entrainment in local minima is dependent on the "annealing schedule," the choice of initial temperature, how many iterations are performed at each temperature, and how much the temperature is decremented at each step as cooling proceeds.

19.2 Simulated Annealing Algorithm Structure

Simulated annealing algorithm can be used for solving various optimization problems [4–15]. A simple example of simulated annealing algorithm is explained through its basic steps according to reference [1].

First, large but finite set v of configurations and the cost C associated with each configuration of set v is defined. The solution of the optimization problem consists of searching the space of configurations for the pair (v, C) presenting the lowest cost. The simulated annealing algorithm starts with an initial configuration v_0 and initial temperature $T = T_0$ and generates a sequence of configurations $N = N_0$. Then, the temperature is decreased after which the new number of steps to be performed is determined according to the temperature level, and the process is then repeated. A candidate configuration is accepted if its cost is less then that of the current configuration. If the cost of the candidate configuration is larger than the cost of the current configuration, it still can be accepted with a certain probability. This ability to perform uphill moves allows simulated annealing to escape from local optimal configurations. The entire process is controlled by a cooling schedule that determines how the temperature decreases during the optimization process.

Figure 19.1 shows the simulated annealing flow chart, which consists of two basic mechanisms: (i) the generation of alternative configurations and (ii) the acceptance rule. The control parameter T_k corresponds to the temperature in physical annealing and N_k is the number of alternatives generated in the kth temperature level. This corresponds with the time that the system stays at a given temperature level and should be large enough for allowing the system to reach a state that corresponds the thermal equilibrium. Initially, when T is large, then larger degradation in the cost function is allowed. As the temperature decreases, the simulated annealing algorithm becomes greedier, and only smaller degradations are accepted. Finally, when T tends to zero, no degradations are accepted.

From the current state S_i, with cost $f(S_i)$, a neighbor solution S_j, with cost $f(S_j)$, is generated by the transition mechanism. Performing the acceptance criteria, the following probability is calculated:

Fig. 19.1 Simulated annealing algorithm flow chart

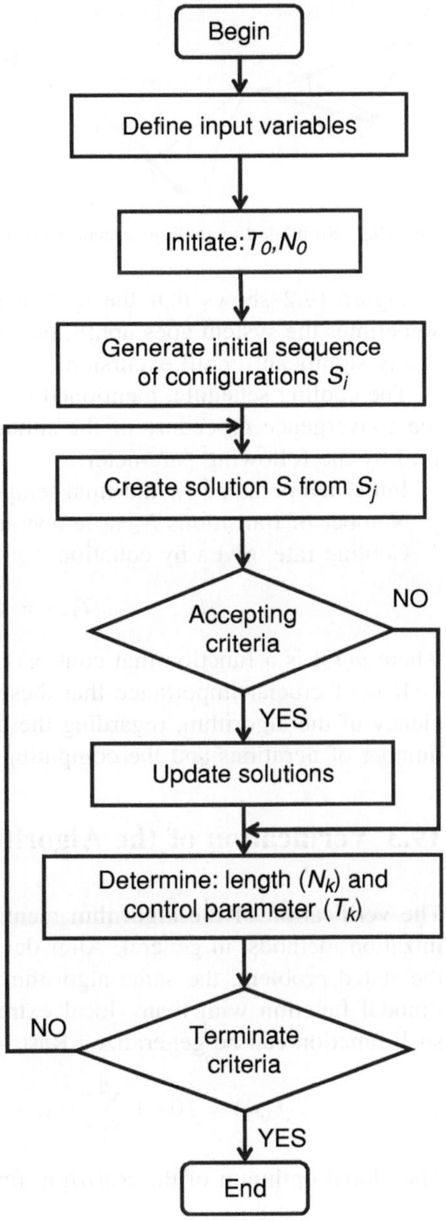

$$p = \begin{cases} 1 & \text{if } f(s_j) \le f(s_i) \\ e^{\frac{f(s_j)-f(s_i)}{T_k}} & \text{if } f(s_j) > f(s_i) \end{cases} \tag{19.2}$$

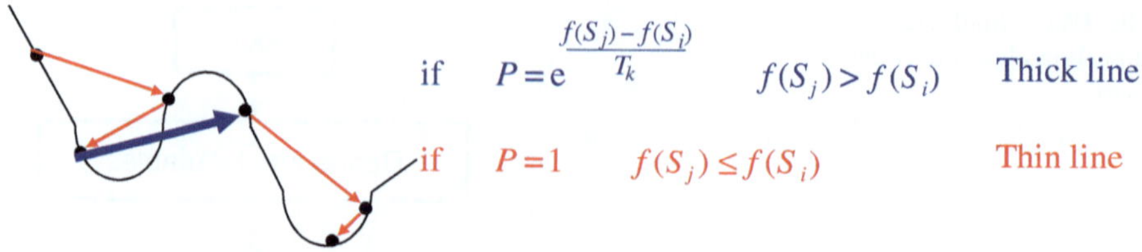

Fig. 19.2 Simulated annealing: escape from local extremum

Figure 19.2 shows that the system goes downhill with probability of 1 and sometimes the system goes uphill, but the lower is the temperature, the less likely is any significant uphill excursion.

The cooling schedule, mentioned above, is the control strategy used to control the convergence procedure of the simulated annealing algorithm. It is characterized by the following parameters:

Initial temperature T_0 and final temperature T_f
Number of transitions N_k at temperature T_k
Cooling rate, given by equation:

$$T_{k+1} = g(T_k) \times T_k \tag{19.3}$$

where $g(T_k)$ is a function that controls the temperature.

It is of crucial importance that these parameters are well chosen for the efficiency of the algorithm, regarding the quality of the final solution as well as the number of iterations and the computing time.

19.3 Verification of the Algorithm

The verification of the algorithm mentioned in this section applies for all optimization methods, in general. After developing the algorithm for optimization of the stated problem, the same algorithm can be examined for solving some multimodal function with many local extrema, which are well known. Example of such function can be generalized Rastrigin function [16].

$$f(x_i) = 10n + \sum_{i=1}^{n} \left[(x_i - x_c)^2 - 10\cos(2\pi(x_i - x_c)) \right] \tag{19.4}$$

The global optimum of the Rastrigin function: $f(x_i) = 0$ at $x_i = x_c$.

19.4 Final Comments

Simulated annealing is a Monte Carlo type of searching technique that randomizes the iterative improvement procedure and also allows occasional uphill moves in attempt to reduce the probability of being stuck at local optimal solution (e.g., local minima). These uphill moves are controlled probabilistically by the temperature *T*, and become less and less likely toward the end of the process, as the value of temperature *T* decreases.

The simulated annealing algorithm is used for solving various optimization problems; thus, it finds its applications in optimization and studies for safe and reliable operation of power systems [6–15].

References

1. Lee KY, El-Sharkawi MA (2008) Modern heuristic optimization techniques: theory and applications to power systems. Wiley, New York
2. Press WH, Flannery B, Teukolsky S, Vettering W (1986) Numerical recipes: the art of scientific computing. Cambridge University Press, Cambridge, UK
3. Metropolis N, Rosenbluth AW, Rosenbluth MN et al (1953) Equation of state calculations by fast computing machines. J Chem Phys 21(6):1087–1092
4. Kirkpatrick S, Gelatt CD, Vecchi MP (1983) Optimization by simulated annealing. Science 220(4598):671–680
5. Bertsimas D, Tsitsiklis J (1993) Simulated annealing. Stat Sci 8(1):10–15
6. Čepin M (2002) Optimization of safety equipment outages improves safety. Rel Eng Syst Saf 77:71–80
7. Mohanta DK, Sadhu PK, Chakrabarti R (2007) Deterministic and stochastic approach for safety and reliability optimization of captive power plant maintenance scheduling using Ga/Sa-based hybrid techniques: a comparison of results. Rel Eng Syst Saf 92:187–199
8. Soliman SA, Mantaway AH, El-Hawary ME (2004) Simulated annealing optimization algorithm for power systems quality analysis. Electric Power Energy Syst 26:31–36
9. Volkanovski A, Čepin M, Mavko B (2009) Nuclear power plant maintenance optimization. In: Briš R, Guedes Soares C, Martorell S (eds) Reliability, risk and safety: theory and applications. Taylor & Francis, London
10. Wong KP (1995) Solving power system optimization problems using simulated annealing. Eng Appl Artif Intel 8(6):665–670
11. Habiballah IO, Irving MR (1995) Multipartitioning of power system state estimation networks using simulated annealing. Electr Pow Syst Res 34:117–120
12. Chen TY, Su JJ (2002) Efficiency improvement of simulated annealing in optimal structural designs. Adv Eng Software 33:675–680
13. Wong SYW (1998) An enhanced simulated annealing approach to unit commitment. Electr Pow Energy Syst 20(5):359–368
14. Basu M (2005) A simulated annealing-based goal-attainment method for economic emission load dispatch of fixed head hydrothermal power systems. Electr Pow Energy Syst 27:147–153
15. Zhu J, Bilbro G, Chow MY (1999) Phase balancing using simulated annealing. IEEE Trans Power Syst 14(4):1508–1513
16. Rastrigin LA (1974) Systems of extremal control. Nauka, Moscow

Part VI
Applications in Practice

Chapter 20
Application of Reliability and Optimization Methods

It is not because things are difficult that we do not dare, it is because we do not dare that they are difficult

Seneca

20.1 Introduction

The field of application of reliability and optimization methods is a broad field covering several theoretical and practical problems and their solutions. The following topics are presented in more details:

- Optimization of test and maintenance intervals of standby equipment
- Reliability analysis of substations
- Configuration control
- Optimization of power plants maintenance schedules
- Optimal generation schedule of power system

20.2 Optimization of Test and Maintenance Intervals of Standby Equipment

Standby equipment is the one that does not operate when the facility is in operation. It is to be placed into operation, if the conditions of the facility require so. For example, onsite power for the switchyard operation comes from the power system. If the blackout occurs, there is an onsite diesel generator in the switchyard, which can provide power for the onsite operations in the switchyard. Diesel generator is standby component in this case. Or, for example, in a nuclear power plant, safety systems are not operating, but they are required to come into operation, if the conditions of the plant require their function to put the plant into the safer state. Each particular safety system is a standby system in this case.

The operability of standby components and systems is demonstrated by the surveillance tests, which are performed periodically every test interval. The test interval is the time between two consecutive moments of starting the test of standby component. If the standby component fails in the period between two

M. Čepin, *Assessment of Power System Reliability*,
DOI: 10.1007/978-0-85729-688-7_20, © Springer-Verlag London Limited 2011

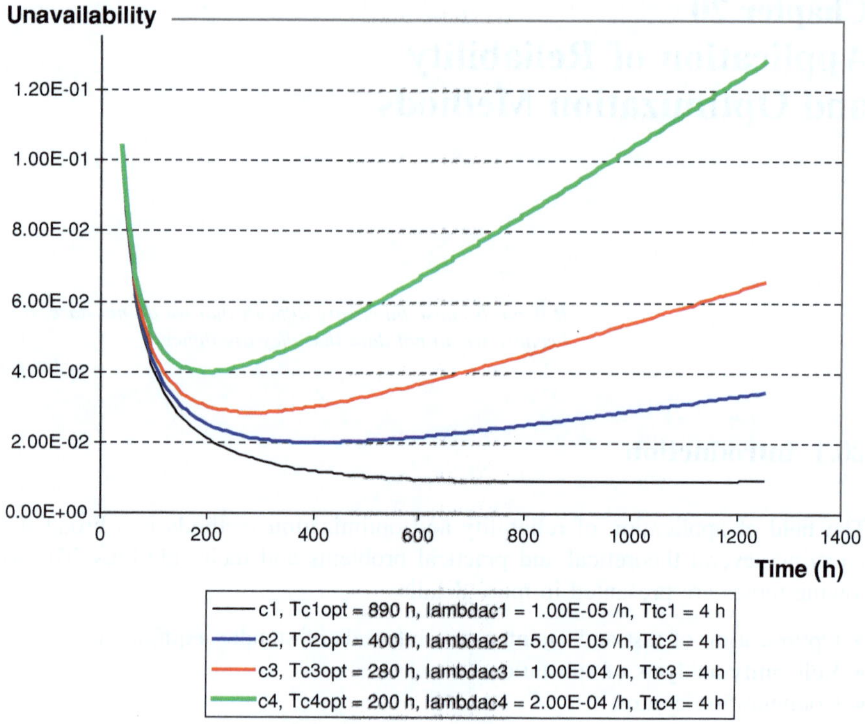

Fig. 20.1 Optimal test interval – family of curves considering constant test duration time

tests, its failure is revealed at the next test. The unavailability of standby component is larger if its failure occurs soon after test. If the moment of failure is a random variable, the unavailability of standby component can be expressed as a function of test interval. If the test time or test duration time, when the standby component may be out of operation because of test, is infinitely small and there are no negative effects of frequent testing, the minimal standby component unavailability would be reached at infinitely short-test interval. In other words, if the standby component would be tested more often, the unavailability would be smaller and smaller, if there are no negative effects of frequent testing.

In the real systems, it can happen that the standby component is switched out of a system for the test duration time, which means that component is unavailable for the test duration time. In real systems, the more frequent testing of particular component may degrade the component more. This means that after test, the component is not as good as it was before it. Several standby components, which may have each its own test interval, may be part of a standby system. The degradation of each of the components may have different influence to the specific test interval of particular component. Several standby systems may exist in the complex facility.

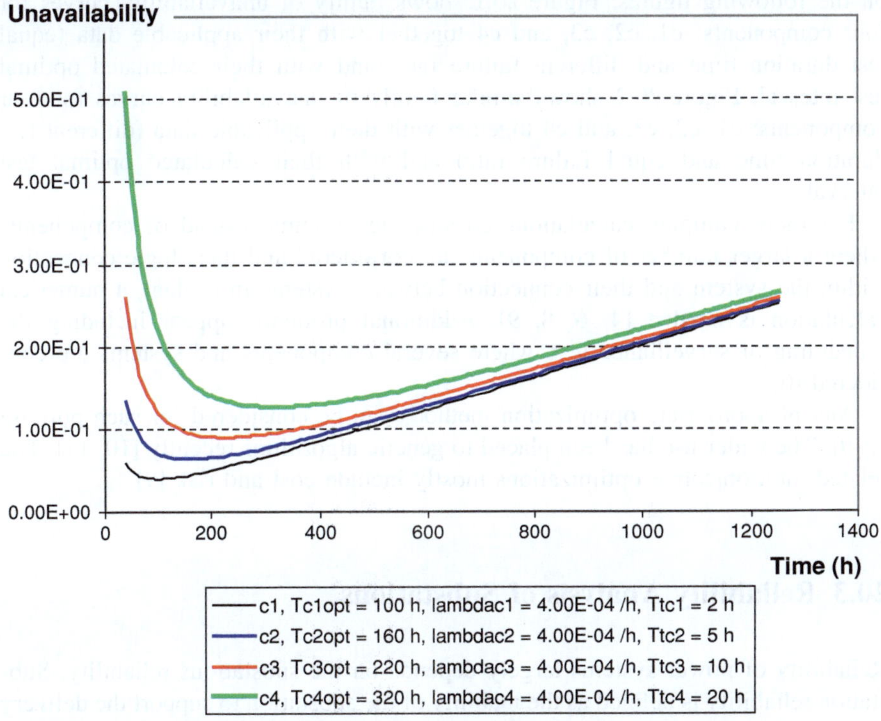

Fig. 20.2 Optimal test interval – family of curves considering constant failure rate

All these facts lead to the conclusion that the mathematical expression of all positive and all negative aspects of surveillance testing of standby equipment is a difficult task. It is a subject of several mathematical models [1–6].

The common characteristic of different models on the level of one component can be the expression of component unavailability as a function of test interval.

$$Q_{\text{mean}}(T_i) = q + \frac{1}{2} \times \lambda T_i + \frac{T_t \times f_o}{T_i} + (q + \lambda \times T_i) \times \frac{T_r}{T_i} \tag{20.1}$$

where Q_{mean} is the mean unavailability, Q is the failure probability, λ is the failure rate, T_i is the test interval, T_t is the test duration time, T_r is the mean time to repair (or mean time to restore), and f_o is the override factor (specifying the portion of test duration time, when component is unavailable).

At point, where the derivative of the unavailability function versus test interval equals 0, the optimum of the function is calculated.

$$\frac{dQ_{\text{mean}}(T_i)}{dT_i} = 0 \Rightarrow T_{\text{iopt}} = \sqrt{\frac{2 \times (T_t \times f_o + q \times T_r)}{\lambda}} \tag{20.2}$$

The resulted optimal test intervals are determined either analytically or numerically or with sensitivity study comparing the families of curves as presented on the following figures. Figure 20.1 shows family of unavailability curves for four components: c1, c2, c3, and c4 together with their applicable data (equal test duration time and different failure rate) and with their calculated optimal test interval. Figure 20.2 shows similar family of unavailability curves for four components: c1, c2, c3, and c4 together with their applicable data (different test duration time and equal failure rate) and with their calculated optimal test interval.

For more complex calculations considering systems instead of components, where a larger number of components are considered and their logic connection within the system and their connection between systems in a plant, a numerical calculation is needed [4, 6, 8, 9]. Additional problems appear including the scheduling of surveillance test, where several components and systems are considered [6].

Any of appropriate optimization methods can be considered for such purpose [7–9]. The wider use has been placed to genetic algorithms recently [10, 11]. The related multiobjective optimizations mostly include cost and risk [9].

20.3 Reliability Analysis of Substations

Reliability of power systems largely depends on the substations reliability. Substation reliability is defined as the inability of the substation to support the delivery of electrical power supply to any of its related outgoing connections or loads.

A reliability model for assessment of substation reliability includes consideration of components, such as circuit breakers, disconnect switches, power lines, and transformers. The developed reliability model can be integrated into the reliability models of power system to evaluate power system reliability statically and to identify the key components for improvement of power system reliability as well as key components to define their maintenance strategy.

The method for assessment of the reliability of power systems can be used for assessment of reliability of substations because of similar configuration features of the system (see Chap. 15) [12–21].

When the reliability analysis is performed, several configurations of substation can be compared in terms of reliability. Not only the configuration of components plays a role in the configuration of the substation but also the normal state of the components is important. Normally closed or normally open circuit breaker can give a significant difference in the calculation of the substation reliability for the same configuration, because the failure mode connected with change of position and failure mode connected with remain in position can be significantly different in terms of their reliability. Namely, passive failures, such as failure to remain in position, have nearly for an order of magnitude smaller failure probabilities compared with active failures, such as failure to change position. Failure mode, failure to change position, can be mostly failure to close although theoretically failure to open suits in this category.

Table 20.1 Example set of data for reliability calculation

Component type	MTTF (year)	MTTR (year)	λ ($year^{-1}$)	μ ($year^{-1}$)	Failure probability
Generator P = 12 MW	3.36E-01	6.85E-03	2.98	1.46E+02	2.00E-02
Generator P = 20 MW	5.14E-02	5.71E-03	19.5	1.75E+02	1.00E-01
Generator P = 50 MW	2.26E-01	2.28E-03	4.42	4.38E+02	1.00E-02
Generator P = 76 MW	2.24E-01	4.57E-03	4.47	2.19E+02	2.00E-02
Generator P = 100 MW	1.37E-01	5.71E-03	7.30	1.75E+02	4.00E-02
Generator P = 155 MW	1.10E-01	4.57E-03	9.13	2.19E+02	4.00E-02
Generator P = 197 MW	1.08E-01	5.71E-03	9.22	1.75E+02	5.00E-02
Generator P = 350 MW	1.31E-01	1.14E-02	7.62	8.76E+01	8.00E-02
Generator P = 400 MW	1.26E-01	1.71E-02	7.96	5.84E+01	1.20E-01
Transformer >550 kV					2.48E-02
Transformer 243–346 kV					1.70E-02
Transformer 146–242 kV					1.61E-02
Transformer 73–145 kV					1.24E-02
Bus 138 kV			1.13E-02	2.09E+02	5.44E-05
Bus 230 kV			9.03E-03	2.04E+02	4.43E-05
Circuit breaker—Active (fails to close)			6.60E-03	8.11E+01	8.14E-05
Circuit breaker—Passive (fails to remain closed)			5.00E-04	8.11E+01	6.16E-06

20.3.1 Reliability Data

More information about good practice and reliability data as a support for the reliability analyses can be found in references [22–31]. Table 20.1 shows an example set of data for reliability calculations. Reliability data for the IEEE-RTS, Reliability Test System, [26] gives the failure rates for lines from 0.02 to 0.54 $year^{-1}$, the repair rates from 11.5 to 876 $year^{-1}$ and the failure probability from 3.88E-04 to 1.75E-03.

The reliability data bases are prepared based on a large and time-consuming processes of collection, classification, and evaluation of failure data [26–29].

20.3.2 Results and Discussion

The results of application of the method to the substations of the test system IEEE-RTS, reliability test system [26], show high reliability of selected substations, which is caused by a selection of low-failure probability data from a set of possible failure data. Table 20.2 shows the results of reliability calculations for selected substations. They are presented in the form of complement of reliability, which is unreliability, for two reasons.

Table 20.2 Results of reliability calculations for selected substations

Substation	Unreliability
3	8.40E-08
4	3.28E-07
7	1.37E-07
8	2.47E-07
9	8.33E-08
13	8.33E-08
16	3.34E-11
20	8.33E-08

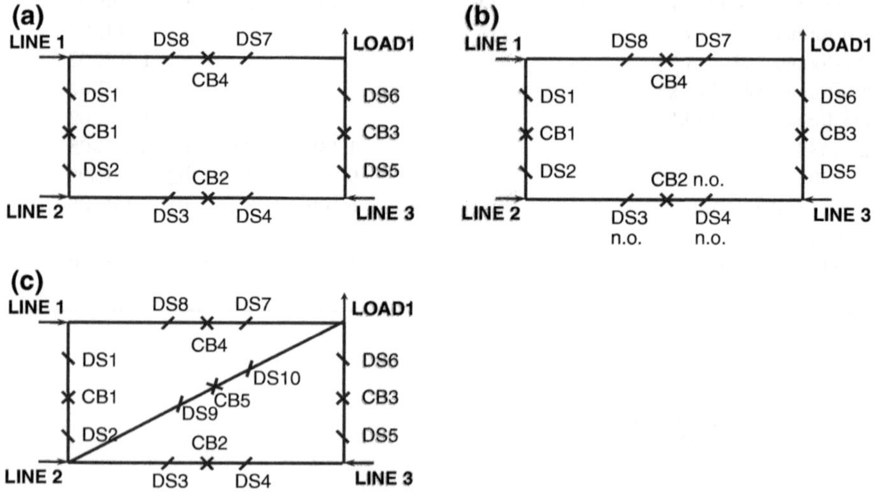

Fig. 20.3 Example comparison of substation configurations

- The reliability calculations are dealt with failure probabilities of components and the result may be consequently expressed in terms of failure probability of a system.
- The results are more representative. Namely, it is much more easier to notice the difference in such small values than in terms of a reliability values, which are very close to one.

Figure 20.3 shows three configurations of a selected substation. Configurations (a) and (b) differ only in normal state of circuit breakers (CB#) and disconnect switches (DS#), which is normally closed if it is not determined as normally open (n.o.). Configuration (c) differs from both regarding additional components.

Several configurations of substation can be compared in terms of reliability and costs [20, 21, 26, 32]. Consideration of costs is not necessarily limited to consideration of investment costs but also to operation costs and test and maintenance costs, for example.

The optimizations of reliability of substations and the optimizations of reliability and costs of substations are not time-consuming optimization assignments. Comparison of all selected configurations can give the optimal solution.

20.4 Configuration Control

Complex facilities with larger number of systems and with larger number of states of particular systems can operate in a large number of configurations. The configurations can be subjected to optimization, because they differ from each other in terms of reliability and safety, consequently.

A good practical example is an outage period in a nuclear power plant where thousands of tests and maintenance activities of safety equipment are performed in relatively short time of several weeks. Even if the plant is in shutdown, some safety functions have to be in place and the minimal number of components and systems has to operate according to the schedule of activities. A large number of test and maintenance activities, which can be in a large extent performed simultaneously, give a tremendous number of plant configurations. A plant configuration is a state of the plant with several systems and components in operation and several systems and components in outage because of test and maintenance activities. The configurations can differ between themselves in risk and reliability measures [33, 34].

Configuration control is the management of component and system arrangements, which differs in component or system status: available versus unavailable, to primarily control risk and reliability. This leads to achieved safety and contributes to the effective use of plant resources.

Configuration control is focused to evaluate timing of placement of testing and maintenance activities. It can be based on optimization, i.e., minimization of system unavailability or accident frequency during a larger period, e.g., complete outage period. Or it can be focused to compare the configurations and to eliminate those configurations, which results in increased risk over the predefined risk limit.

20.5 Optimization of Power Plants Maintenance Schedules

Maintenance of power plants is one of important actions connected to the overall power system reliability. Maintenance is done mostly based on consideration of the power-generating units of the power plants as the basic elements of the maintenance. Maintenance of the power-generating units is a timely process, which means that the generating units are periodically taken out of operation and are subjected to the maintenance activities [35–52].

The maintenance schedule, which is not optimized, reduces system reliability and increases the costs of the electric power system. Its affects many short-term, middle-term, and long-term planning functions, including unit commitment, fuel scheduling, optimal use of water resources, long-term power system development planning, reliability calculations, and costs. If the maintenance schedule is not optimized, those functions can be adversely affected.

When optimized, the maintenance schedule increases the power system reliability, reduces costs, and extends the generator lifetime.

20.5.1 Mathematical Model and Procedure

The maintenance scheduling of power-generating units in a power system is a complex combinatorial optimization problem with non-linear objectives and constraints [35]. The annual value of loss of load expectation (LOLE) of power system is an objective function requiring minimization in the analyzed period.

$$\min \mathrm{LOLE}_a(X_{\mathrm{NG}}) \tag{20.3}$$

where $\mathrm{LOLE_a}$ is the a calculated annual loss of load expectation in days/year for X_{NG} maintenance schedules of NG generators, X_{NG} is the maintenance schedule of unit X given as day/week when maintenance of unit X starts, and NG is the number of generating units in power system.

The number of the independent variables in a model is determined by the number of units in the power system. The objective function is subjected to operating and planning constraints.

The evaluation of LOLE for a particular configuration of generating units consists of three parts: (i) the load model, (ii) the capacity or generation model, and (iii) the risk model. The capacity model is formed with direct analytical methods by creating a capacity outage probability table. This table represents the capacity outage states of the generating units together with the probabilities of each state. The load model can either be the daily peak load variation curve, which only includes the peak loads of each day, or the load duration curve, which represents the hourly variation of the load. The risk indice, i.e., loss of load expectation, is evaluated by convolution of the capacity and load model, obtaining the expected number of days in the specified period in which the daily peak load exceeds the available capacity (LOLE_p):

$$\mathrm{LOLE}_p = \sum_{i=1}^{n} P_i(C_i - L_i) \quad \text{days/period} \tag{20.4}$$

where $\mathrm{LOLE_p}$ is the value of LOLE in period p with n days, C_i is the available capacity on day i, L_i is the forecast peak load on day i, $P_i(C_i - L_i)$ is the probability of loss of load on day i, obtained directly from the capacity outage cumulative probability table.

Loss of load expectation method provides a consistent and sensitive measure of power-generating system reliability.

The value of weekly LOLE_p is calculated using previous equation and $n = 7$ and those values are summed to calculate the annual LOLE_a:

$$\mathrm{LOLE}_a = \sum_{p=1}^{52} \mathrm{LOLE}_p \tag{20.5}$$

The daily peak load variation curve is used to evaluate LOLE giving a risk expressed in number of days during the period of study in which the load exceeds available capacity. The calculated indices measure the overall adequacy of the power-generating system to cover total system load without taking into account separate load points, transmission constraints, or energy available in the system.

Load forecast is normally predicted on historical load data, and its uncertainty can be described by a probability distribution whose parameters can be determined from past experience, load forecast model, or possible subjective evaluation. The uncertainty of load forecasting can be included in the risk computations by dividing the load forecast probability distribution into class intervals mid-value. The LOLE is computed for each load represented by the class interval and multiplied by the probability of existence of the load. The sum of these products represents the LOLE for the forecast load. The load forecast uncertainty is the single most important parameter in operating capacity reliability evaluation in addition to the unit reliability.

The probability of finding the generating unit in forced outage at some distant time in the future represents the basic parameter used in static capacity evaluation for modeling generating unit reliability. This probability is defined as unit unavailability, which is in power system applications known as the unit-forced outage rate (see Chap. 13).

$P_0(t)$ and $P_1(t)$ are the probabilities of being found in the operating state and failed state, respectively, as a function of time given that the system started at time $t = 0$ in the operating state.

$$p_0(t) = \frac{\mu}{\lambda + \mu} + \frac{\lambda e^{-(\lambda+\mu)t}}{\lambda + \mu} \tag{20.6}$$

$$p_1(t) = \frac{\mu}{\lambda + \mu} + \frac{\lambda e^{-(\lambda+\mu)t}}{\lambda + \mu} \tag{20.7}$$

The probability of being found in the failed state for time-approaching infinity equals the unavailability.

$$p_1(\infty) = U = \mathrm{FOR} = \frac{\lambda}{\lambda + \mu} \tag{20.8}$$

Capacity outage probability table is calculated (see Chap. 13).

The constraints are determined. The capacity constraint is modeled by the limiting maximum value of calculated period LOLE_p. The additional constraints can be considered: the network overload or other contingencies connected with the power transfer capacity of transmission network.

The optimization method is selected and applied (see chapters on specific selected optimization method). The details of applying genetic algorithms are in reference [35].

The results show the timing of the maintenance of each generating unit. Typically, the outage of larger units is suggested at the periods of smaller consumption.

20.6 Optimal Generation Schedule of Power System

The economic dispatch is one of the most important tasks of power system operation and planning [54, 55]. The purpose of the economic dispatch is to schedule the outputs of all available generation units in the power system to minimize the fuel cost while satisfying all unit and system equality and inequality constraints including those connected to minimization of the emission of gaseous pollutants [53–68]. When minimization of the emission of gaseous pollutants is considered, the economic dispatch is denoted as economic environmental power dispatch.

The network security constraints are important in practice, so it is very important to solve the economic dispatch with their consideration [69]. This section briefly summarizes the economic environmental power dispatch. Its extension considering the network security constraints including N security economic dispatch and $N-1$ security economic dispatch is well presented in reference [69] in theory and examples and it is not repeated here.

20.6.1 Mathematical Model

The fuel cost for thermal-generating unit is determined by second-order function of the active power generation as follows:

$$FC_i(P_{Gi}) = a_i + b_i \times P_{Gi} + c_i \times P_{Gi} \tag{20.9}$$

where a_i (\$/h), b_i (\$/MWh), and c_i (\$/MW^{2}h) are the constants which are unique for each generating unit, P_{Gi} is the power output of the ith thermal unit, and FC_i is the fuel cost of thermal unit i.

An example of costs is summarized in reference [53]. The total fuel cost from all thermal units in the system during the studied time period is determined as the sum of all costs.

$$f_c = \sum_{t=1}^{T} \sum_{i=1}^{I} FC_i(P_{Git}) \tag{20.10}$$

where P_{Git} is the power output of the ith thermal unit at time interval t, T is the number of considered time intervals under study, and I is the number of thermal units.

The total emission of atmospheric pollutants, such as sulfur oxide and nitrogen oxide, caused by thermal unit can be modeled either separately or jointly. If the total emission of atmospheric pollutants is modeled jointly, the following equation is applicable.

$$\mathrm{FE}_i(P_{Gi}) = \alpha_i + \beta_i \times P_{Gi} + \gamma_i \times P_{Gi}^2 + \eta_i \times e^{\lambda_i P_{Gi}} \tag{20.11}$$

where α_i (t/h), β_i (t/MWh), γ_i (t/MW^{2}h), η_i (t/h), and λ_i (1/MW) are the constants that are unique for each generating unit.

The total emission of atmospheric pollutants from all thermal units in the system during the studied time period is determined as the sum of all emissions.

$$f_E = \sum_{t=1}^{T}\sum_{i=1}^{I} \mathrm{FE}_i(P_{Git}) \tag{20.12}$$

The coefficients $\alpha_i, \beta_i, \gamma_i, \mu_I, \lambda_i$ are equal to zero for nuclear power units.

The power balance equations constraint describes the fact that the sum of output powers of all generating units must equal to the total load demand at each time interval.

$$P_{Lt} = \sum_{i=1}^{I} P_{Git} + \sum_{j=1}^{J} P_{Hjt} \quad t = 1, 2, \ldots, T \tag{20.13}$$

where J is the number of reservoirs, P_{Hjt} is the water to power conversion function of the power plant associated with reservoir j is determined as follows [65, 66]:

$$P_{Hjt} = C_{ij} \times V_{jt}^2 + C_{2j} \times X_{jt}^2 + C_{3j} \times V_{jt} \times X_{jt} + C_{4j} \times V_{jt} + C_{5j} \times X_{jt} + C_{6j} \tag{20.14}$$

where $C_{i,j}$ is the hydro ith power generation coefficient for jth reservoir, V_{jt} is the storage volume for jth reservoir at time interval t in m^3, and X_{jt} is the volume of water discharged from jth reservoir during time interval t in m^3.

The generator capacity constraints are expressed considering the minimal and maximal power output for each unit.

$$P_{Gi}^{\min} \le P_{Git} \le P_{Gi}^{\min} \tag{20.15}$$

$$P_{Hj}^{\min} \le P_{Hjt} \le P_{Hj}^{\min} \tag{20.16}$$

$P_{Gi}^{\min}$ and $P_{Gi}^{\max}$	Minimum and maximum power output for ith thermal unit, respectively
$P_{Hj}^{\min}$ and $P_{Hj}^{\max}$	Minimum and maximum power output for jth hydro unit, respectively

The power produced by each thermal unit is bounded by its ramping response rate limits [64, 67].

$$-\mathrm{MI}_i \leq P_{Git} - P_{Git+1} \leq \mathrm{MD}_i \tag{20.17}$$

where MI_i is the maximum increase in the output of the ith thermal unit over one time interval, and MD_i is the maximum decrease in the output of the ith thermal unit over one time interval.

Hydraulic constraints include limitations about water volume in reservoirs of hydro power plants. Water volume in each reservoir j at each time interval t is limited.

$$V_{jt+1} = V_{jt} - X_{jt} + I_{jt} \qquad t = 0, 1, 2, \ldots, T - 1 \tag{20.18}$$

where I_{jt} is the inflow volume in the jth reservoir during time interval t.

Physical volume limitations of each reservoir in storage are limited.

$$V_j^{\min} \leq V_{jt} \leq V_j^{\max} \tag{20.19}$$

$$X_j^{\min} \leq X_{jt} \leq X_j^{\max} \tag{20.20}$$

where $V_j^{\min}$and $V_j^{\max}$ are the minimum and maximum water volume for jth reservoir, respectively, and $X_j^{\min}$and $X_j^{\max}$ are the minimum and maximum water volume discharge from the jth reservoir, respectively.

The initial and the final reservoir storage volumes are determined.

$$V_{jt}\big|^{t=0} = V_j^{\text{begin}} \tag{20.21}$$

$$V_{jt}\big|^{t=T} = V_j^{\text{end}} \tag{20.22}$$

where $V_j^{\text{begin}}$and V_j^{end} are the initial and final water volume in the jth reservoir, respectively.

Functions f_C and f_E are the objective functions, i.e., the functions that have to be minimized. Aggregating the objectives and constraints, the economic environmental power dispatch optimization problem can be mathematically formulated as a nonlinear constrained multi-objective optimization problem.

$$\text{minimize}[f_c(P_{Git}), f_E(P_{Git})] \tag{20.23}$$

Subjected to equality and inequality constraints:

$$g(x) = 0 \tag{20.24}$$

$$g(x) \leq 0 \tag{20.25}$$

where x is the decision vector which represents possible solution.

The multi-objective economic environmental power dispatch optimization problem minimizing functions f_C and f_E can be transformed into the single objective optimization problem.

$$\text{minimize} \quad f = w \times f_c + (1 - w) \times \sigma \times f_E \tag{20.26}$$

where σ is the scaling factor, and w is the weighting factor which is varied.

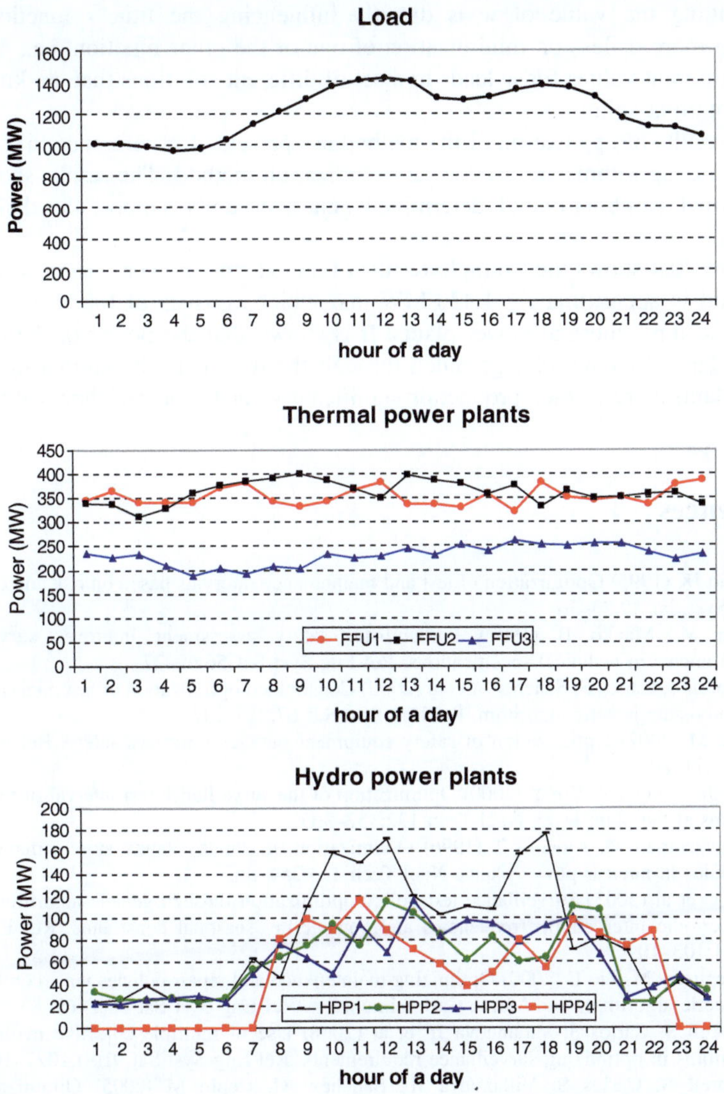

Fig. 20.4 Example case including thermal and hydro power plants supplying the load

The scaling factor is applied to approximately equalize the values of both functions, the cost function f_C and the emission function f_E, giving them equal chances to be minimized. The scaling factor is determined as the ratio between the maximum fuel cost and maximum emission.

$$\sigma = \frac{f_c(P_{Gi}^{\max})}{f_E(P_{Gi}^{\max})} \tag{20.27}$$

Changing the value of w is directly influencing the fitness function, i.e., focusing more or less on minimization of one or the other function, i.e., f_C or f_E. Set of different values for w leads to a set of different solutions that are known as Pareto optimal solutions.

The results of application of the method are presented in reference [68], where the genetic algorithms are used as an optimization method. The results show how the optimal combination of determined generator units support the daily load diagram.

Figure 20.4 shows the example results of an example power system, where the peak load is approximately 1,420 MW and which consists of four hydro power plants and three thermal power plants. It is shown that the power of the thermal power plants does not change much through the day. It is shown that the hydro power plants increase their production significantly during the peak hours of the day.

References

1. Vaurio JK (1995) Optimization of test and maintenance intervals based on risk and cost. Rel Eng Syst Saf 49:23–36
2. Čepin M, Mavko B (1997) Probabilistic safety assessment improves surveillance requirements in technical specifications. Rel Eng Syst Saf 56:69–77
3. Martorell S, Carlos S, Sanchez A et al (2000) Constrained optimization of test intervals using a steady-state genetic algorithm. Rel Eng Syst Saf 67:215–232
4. Čepin M (2002) Optimization of safety equipment outages improves safety. Rel Eng Syst Saf 77:71–80
5. Yang JE, Sung TY, Yin Y (2000) Optimization of the surveillance test interval of the safety systems at the plant level. Nucl Tech 132:352–365
6. Harunuzzaman M, Aldemir T (1996) Optimization of standby safety system maintenance schedules in nuclear power plants. Nucl Tech 113:354–367
7. Lapa C et al (2003) Surveillance test policy optimization through genetic algorithms using non-periodic intervention frequencies and considering seasonal constraints. Rel Eng Syst Saf 81:103–109
8. Marseguerra M, Zio E (2000) Optimizing maintenance and repair policies via a combination of genetic algorithms and Monte Carlo simulation. Rel Eng Syst Saf 68(1):69–83
9. Martorell S, Carlos S, Villanueva JF et al (2006) Use of multiple objective evolutionary algorithms in optimizing surveillance requirements. Rel Eng Syst Saf 91(9):1027–1038
10. Martorell S, Carlos S, Villanueva JF, Sánchez AI, Čepin M (2005) Optimization of surveillance requirements at NPP considering time-dependent scheduling of tests and preventive maintenance. In: Proceedings of the European and reliability conference, ESREL, Balkema, Taylor & Francis, pp 1363–1370
11. Martorell S, Sánchez A, Carlos S et al (2004) Alternatives and challenges in optimizing industrial safety using genetic algorithms. Rel Eng Syst Saf 86(1):25–38
12. Roberts NH, Vesely WE, Haasl DF, Goldberg FF (1981) Fault tree handbook. NUREG-0492 NRC, Washington
13. Vesely W, Dugan J, Fragola J et al (2002) Fault tree handbook with aerospace applications. NASA
14. Čepin M, Mavko B (2002) A dynamic fault tree. Rel Eng Syst Saf 75(1):83–91
15. Čepin M (2005) Analysis of truncation limit in probabilistic safety assessment. Rel Eng Syst Saf 87(3):395–403

16. Mosleh A, Fleming KN (1988) Procedures for treating common cause failures in safety and reliability studies. NUREG/CR-4780, NRC, Washington
17. Volkanovski A, Čepin M, Mavko B (2009) Application of the fault tree analysis for assessment of power system reliability. Rel Eng Syst Saf 94(6):1116–1127
18. Volkanovski A, Čepin M, Mavko B (2008) Slovenian power system reliability analysis. In: TopSafe 2008 Proceedings international topical meeting on safety of nuclear installations. Dubrovnik, ENS
19. Kančev D, Čauševski A, Čepin M, Volkanovski A (2007) Application of probabilistic safety assessment for macedonian electric power system. In: Proceedings of international conference on nuclear energy for new Europe, NSS
20. Čepin M (2010) Switchyard fault tree analysis: reliability, risk and safety: back to the future. In: Proceedings ESREL2010, Taylor & Francis, pp 638–643
21. Čepin M (2010) Assessment of switchyard reliability with the fault tree analysis. In: Proceedings of international conference on probabilistic safety assessment and management, IAPSAM
22. IEEE Gold Book (2007) IEEE recommended practice for the design of reliable industrial and commercial power system, ANSI/IEEE Std 493-2007, IEEE
23. IEEE Standard 500 (1984) IEEE guide to the collection and presentation of electrical, electronic, sensing component, and mechanical equipment reliability data for nuclear-power generating stations, appendix D, reliability data for nuclear-power generating stations, IEEE
24. T-Book: reliability data of components in Nordic nuclear power plants (2000) Villingby, Sweden TUD Office and Pörn Consulting
25. IAEA-TECDOC-478 (1988) Component reliability data for use in probabilistic safety assessment. IAEA
26. Grigg C, Wong P, Albrecht P, Allan R, Bhavaraju M, Billinton R, Chen Q, Fong C, Haddad S, Kuruganty S, Li W, Mukerji R, Patton D, Rau N, Reppen D, Schneider A, Shahidehpour M, Singh C (1999) The IEEE reliability test system—1996: a report prepared by the Reliability Test System Task Force of the application of probability methods subcommittee. IEEE Trans Power Syst 14(3):1010–1020
27. Atwood CL, La Chance JL, Martz HF et al (2003) Handbook of parameter estimation for probabilistic risk assessment. NUREG/CR-6823, NRC
28. Denson W, Crowell WH, Mahar D et al (1996) Electronics parts reliability data: a compendium of commercial and military device field failure rates, 1997, EPRD-97. Reliability Analysis Center
29. Drago JP, Borkowski RJ, Pike DH et al (1982) The in-plant reliability data base for nuclear-power-plant components: data collection and methodology report, NUREG/CR-2641. Oak Ridge National Laboratory, NRC
30. Reliability Prediction of Electronic Equipment (1986) Issue E, MIL-HDBK217E, Department of Defense
31. European Industry Reliability Data Handbook (1991) Electrical Power Plants EUROSTAT, EIREDA, Paris
32. Volkanovski A (2008) Impact of offsite power system reliability on nuclear power plant safety. PhD thesis, University of Ljubljana
33. Kiper KL (2002) Insights from an all-modes PSA at Seabrook Station. In: Proceedings of the international topical meeting on probabilistic safety assessment, Detroit, ANS, pp 429–434
34. NUREG/CR-6144 (1995) Evaluation of potential severe accident during low power and shutdown operations at Surry, Unit 1, NRC
35. Volkanovski A, Mavko B, Boševski T et al (2008) Genetic algorithm optimization of the maintenance scheduling of generating units in a power system. Rel Eng Syst Saf 93:779–789
36. Volkanovski A (2005) Optimization of the maintenance of the generating units in the power system using genetic algorithm. Master's thesis, Faculty of Electrical Engineering, Skopje
37. Dopazo JF, Merrill HM (1975) Optimal generator maintenance scheduling using integer programming. IEEE Trans Power Appar Syst 94(5):1537–1545

38. Yamayee ZA (1982) Maintenance scheduling: description, literature survey and interface with overall operations scheduling. IEEE Trans Power Appar Syst 101(8):2770–2779
39. Podbury C, Dillon T (1987) An intelligent knowledge based system for maintenance scheduling in a power system. In: Proceedings of the 9th power systems computation conference (PSCC), Lisbon, pp 708–714
40. Choueiry B, Sekine Y (1988) Knowledge based method for power generators maintenance scheduling. In: Proceedings of the expert systems applications to power systems (ESAP), Helsinki, pp 97–14
41. Chen L, Toyoda J (1990) Maintenance scheduling based on two-level hierarchical structure to equalize incremental risk. IEEE Trans Power Syst 5:1510–1516
42. Satoh T, Nara K (1991) Maintenance scheduling by using simulated annealing. IEEE Trans Power Syst 6(2):850–857
43. Chen L, Toyoda J (1991) Optimal generating units maintenance scheduling for multi-area system with network constraints. IEEE Trans Power Syst 6:68–74
44. Mukerji R, Merrill H, Ericson B et al (1991) Power plant maintenance scheduling-optimizing economics and reliability. IEEE Trans Power Syst 6:476–483
45. Yellen J, Al-Khamis TM, S Vemuri et al (1992) A decomposition approach to unit maintenance scheduling. IEEE Trans Power Appar Syst 7(2):726–733
46. Al-Khamis TM, Yellen J, Vemuri S et al (1992) Unit maintenance scheduling with fuel constraints. IEEE Trans Power Appar Syst 7(2):933–939
47. Lin C, Huang C, Lee S (1992) An optimal generators maintenance approach using fuzzy dynamic programming. In: Proceedings of IEEE/PES summer meeting, Seattle, pp 401–410
48. Kralj B, Rajakovic N (1994) Multiobjective programming in power system optimization: new approach to generator maintenance scheduling. Int J Electr Power Energy Syst 16(4):24–32
49. Dahal K, McDonald J (1997) A review of generator maintenance scheduling using artificial intelligence techniques. In: Proceedings of the 32nd University Power Eng Conf (UPEC), Manchester, pp 787–790
50. Kim H, Hayashi Y, Nara K (1997) An algorithm for thermal unit maintenance scheduling through combined use of GA, SA and TS. IEEE Trans Power Syst 12:329–335
51. Bretthaner G, Gamaleja T, Handschin E et al (1998) Integrated maintenance scheduling system for electrical energy systems. IEEE Trans Power Delivery 132:655–660
52. Moro L, Ramos A (1999) Goal programming approach to maintenance scheduling of generating units in large scale power system. IEEE Trans Power Syst 14:1021–1028
53. Kaplan S (2008) Power plants: characteristics and costs. CRS Report for Congress
54. Wood AJ, Wollenberg BF (1996) Power generation operation and control. Wiley, New York
55. Grainger JJ, Stevenson WD (1994) Power system analyses. McGraw-Hill, London
56. King TD, El-Hawary ME, El-Hawary F (1995) Optimal environmental dispatching of electric power systems via an improved hopfield neural network model. IEEE Trans Power Syst 10(3):1559–1565
57. Zhang PX, Zhao B, Cao YJ et al (2004) A novel multi-objective genetic algorithm for economic power dispatch. IEEE Universities Power Eng Conf 1:422–426
58. Bharathi R, Kumar MJ, Sunitha D et al (2007) Optimization of combined economic and emission Dispatch problem: a comparative study. IEEE Power Eng Conf 134–139
59. Ozyon S, Yasar C, Aslan Y et al (2009) Solution to environmental economic power dispatch problems in hydrothermal power systems by using genetic algorithm. IEEE Conf Electr Electron Eng I-387–I-391
60. Abido MA (2009) Multiobjective particle swarm optimization for environmental economic dispatch problem. Electr Pow Syst Res 79:1105–1113
61. Palanichama C, Babu NS (2008) Analytical solution for combined economic and emissions dispatch. Electr Pow Syst Res 78:1129–1137
62. Lakshminarasimman L, Subramanian S (2008) A modified hybrid differential evolution for short-term scheduling of hydrothermal power systems with cascaded reservoirs. Energ Convers Manag 49:2513–2521

63. Mandal KK, Chakraborty N (2008) Differential evolution technique-based short-term economic generation scheduling of hydrothermal systems. Electr Pow Syst Res 78:1972–1979
64. Simopoulos DN, Kavatza SD, Vournas CD (2007) An enhanced peak shaving method for short term hydrothermal scheduling. Energ Convers Manag 48:3018–3024
65. Basu M (2008) Dynamic economic emission dispatch using nondominated sorting genetic algorithm-II. Electr Pow Energ Syst 30:140–149
66. Liang RH, Liao JH (2007) A fuzzy-optimization approach for generation scheduling with wind and solar energy systems. IEEE Trans Power Syst 22(4):1665–1674
67. Haupt RL, Haupt SE (2003) Practical genetic algorithms. Wiley, New York
68. Gjorgiev B (2010) Fuzzy-genetic optimization approach for generation scheduling with system consisted of conventional and renewable energy sources. Master's thesis, 2010
69. Zhu J (2009) Optimization of power system operation. Wiley, Piscataway, NJ

Index

M. Čepin, *Assessment of Power System Reliability*,
DOI: 10.1007/978-0-85729-688-7, © Springer-Verlag London Limited 2011

MIX
Papier aus verantwortungsvollen Quellen
Paper from responsible sources
FSC® C105338

If you have any concerns about our products,
you can contact us on
ProductSafety@springernature.com

In case Publisher is established outside the EU,
the EU authorized representative is:
Springer Nature Customer Service Center GmbH
Europaplatz 3, 69115 Heidelberg, Germany

Printed by Libri Plureos GmbH
in Hamburg, Germany